The Antarctic Treaty regime:
law, environment and resources

Studies in Polar Research

This series of publications reflects the growth of research activity in and about the polar regions, and provides a means of disseminating the results. Coverage is international and interdisciplinary: the books will be relatively short (about 200 pages), but fully illustrated. Most will be surveys of the present state of knowledge in a given subject rather than research reports, conference proceedings or collected papers. The scope of the series is wide and will include studies in all the biological, physical and social sciences.

Other titles in this series:
The Antarctic Circumpolar Ocean
Sir George Deacon
The Living Tundra
Yu. I. Chernov, transl. D. Love
Transit Management in the Northwest Passage
edited by C. Lamson and D. Vanderzwaag
Arctic Air Pollution
edited by B. Stonehouse

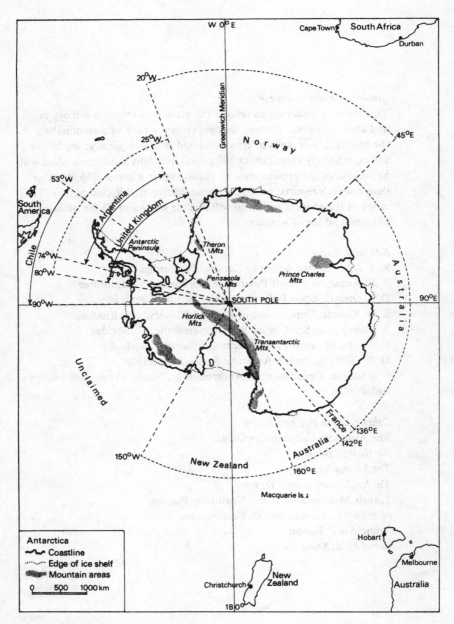

Map of national claims in Antarctica

The Antarctic Treaty regime

LAW, ENVIRONMENT AND RESOURCES

Edited by

GILLIAN D. TRIGGS

Senior Lecturer, Law School, University of Melbourne

Under the General Editorship of the
British Institute of International and Comparative Law

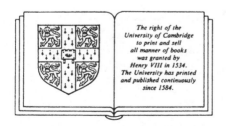

The right of the
University of Cambridge
to print and sell
all manner of books
was granted by
Henry VIII in 1534.
The University has printed
and published continuously
since 1584.

CAMBRIDGE UNIVERSITY PRESS
Cambridge
New York   Port Chester   Melbourne   Sydney

CAMBRIDGE UNIVERSITY PRESS
Cambridge, New York, Melbourne, Madrid, Cape Town, Singapore, São Paulo, Delhi

Cambridge University Press
The Edinburgh Building, Cambridge CB2 8RU, UK

Published in the United States of America by Cambridge University Press, New York

www.cambridge.org
Information on this title: www.cambridge.org/9780521327664

First published 1987
Reprinted 1989
This digitally printed version 2008

A catalogue record for this publication is available from the British Library

Library of Congress Cataloguing in Publication data
The Antarctic Treaty regime.
   Papers from the proceedings of a conference organised by the British Institute
of International and Comparative Law, in London, April 1985.
   Bibliography
   Includes index.
   1. Antarctic regions – International status – Congresses.   2. Marine pollution
– Law and legislation – Antarctic regions – Congresses.   3. Mining Law – Ant-
arctic regions – Congresses.   I. Triggs, Gillian D. (Gillian Doreen), 1945–
II. British Institute of International and Comparative Law.
JX4084.A5A556 1987   341.2′9′09989   86-21564

ISBN 978-0-521-32766-4 hardback
ISBN 978-0-521-10008-3 paperback

Contents

Contributors

James H. Barnes, *Director, The Antarctica Project, Washington DC.*

William N. Bonner, *Head, Life Sciences Division, British Antarctic Survey, UK.*

David J. Drewry, *Director, Scott Polar Research Institute, University of Cambridge, UK.*

Hazel Fox, *Director, British Institute of International and Comparative Law, London, UK.*

John A. Gulland, *Imperial College, London and FAO.*

John A. Heap, *Head of Polar Regions Section, South America Department, Foreign and Commonwealth Office, UK.*

Martin W. Holdgate, *Chief Scientist, Department of Environmental and Development of Transport, UK.*

F. G. Larminie, *Coordinator – External Affairs, Health, Safety and Environmental Services for British Petroleum, UK.*

Richard M. Laws, *Director, British Antarctic Survey, Cambridge, UK.*

John Rowland, *former Ambassador of Australia, Department of Foreign Affairs.*

Gillian D. Triggs, *Senior Lecturer, Law School, University of Melbourne, Australia.*

Rolph Trolle-Andersen, *Minister Plenipotentiary, Royal Ministry of Foreign Affairs, Norway.*

Francisco Orrego Vicuña, *Director of the Institute of International Studies of the University of Chile, Santiago, Chile.*

A. D. Watts, *Deputy Legal Advisor, Foreign and Commonwealth Office, London, UK.*

H. E. Ambassador Zain-Azraai, *Permanent Representative of Malaysia to the United Nations.*

Antarctic Treaty Parties

Total: 32 states.

12 original treaty signatories 1959

7 claimant states

Argentina	United Kingdom
Australia	United States
Chile	Japan
France	U.S.S.R.
New Zealand	South Africa
Norway	Belgium

18 Consultative Parties including the first listed 6 acceding states

20 acceding states

1961 Poland	1974 German Democratic
1979 Federal Republic of	Republic
Germany	1984 Hungary
1975 Brazil	1981 Italy
1983 India	1967 Netherlands
1980 Uruguay	1981 P.N.G.
1983 Peoples' Republic of	1981 Peru
China	1971 Romania
1978 Bulgaria	1982 Spain
1962 Czechoslovakia	1984 Sweden
1965 Denmark	1984 Cuba
1984 Finland	

Foreword

The Antarctic is unique in so many respects that is difficult to avoid superlatives in any description. Flying up the Beardmore Glacier from the New Zealand headquarters in McMurdo Sound at Scott Base in a US Navy Hercules C130 I was struck by the marvellous beauty of the Antarctic and yet its innate harshness. Landing at the South Pole under the midnight sun and spending a brief few hours at the American base one at once realises how different it all is to the rest of the world, including even north polar regions. While the North Pole is situated on sea ice over an ocean 3000 m deep, the South Pole is in the middle of a plateau of ice of similar depth, stretching as it seems endlessly in all directions.

There are great mountains distant from the Pole, and amongst those nearer the coast are impressive dry valleys and canyons, empty even of snow and sometimes with a small lake which very occasionally thaws. At the New Zealand base at Lake Vanda the temperature at the bottom of the lake is −25 °C. Nearby there is a small strip of brown moss close to the glaciers. In the few ice-free areas have been found not only coal deposits, but also traces of minerals which could be of economic interest. It will be many years, if ever, before these will be exploited, though it is perhaps otherwise with the potential hydrocarbon fields under the Antarctic seas, not to mention the enormous stocks of fish, squid and krill in the ocean south of the Antarctic Convergence.

It is against this background – the fear of some future scramble for Antarctic resources – that the Antarctic Treaty, one of the most successful treaties ever to have been negotiated, will need to be further developed. Under the present Treaty, all the territorial claims in the Antarctic, some of which are overlapping, are literally frozen indefinitely. In the Antarctic continent the 'Cold War' does not exist. It is 'nuclear free' and all bases of all countries are open to inspection. The Antarctic is still a 'continent for science'.

The importance of this scientific work is undoubted, and the cooperation of world scientists is excellent. Indeed, there is science, some of immediate relevance, particularly the ionospheric work, which can be more effectively done in the Antarctic than anywhere else. The climatic history of the world over thousands of years is found frozen into the Antarctic ice. Thirty-two nations are parties to the Antarctic Treaty and amongst the eighteen full members who are actually carrying out scientific work in the Antarctic are countries as diverse as Argentina, India, Poland, the United States, the Soviet Union, China and, of course, Britain, which is playing a leading scientific role.

A major concern of Antarctic scientists is the conservation of this still largely unpolluted continent which could, however, be threatened if the Antarctic Treaty fails to develop. Indeed, the Antarctic Treaty could be at risk as it stands. While there is no time limit, it will be open for any nation in the year 1991, or thereafter, to seek amendments to the Treaty or, after an interval to withdraw. Therefore it is right that lawyers should turn their earnest attention to consider the future regime for the Antarctic.

All these issues and the geopolitical implications were discussed at a conference organised by the British Institute of International and Comparative Law in London in April 1985, admirably organised by their Director Hazel Fox, and owing a great deal to Dr Gillian Triggs of Melbourne University. Not only were there leading lawyers present, such as Dr Francisco Orrego, who at that time was Chilean Ambassador in London, and Arthur Watts, Legal Counsellor at the Foreign and Commonwealth Office (FCO), but also many distinguished scientists and public servants.

The contributors to this volume are drawn from many disciplines and many countries, and represent many viewpoints. Here we have contributions from leading scientists: Dr R. M. Laws of the British Antarctic Survey; Dr Martin Holdgate, Chief Scientist and Deputy Secretary, Department of the Environment, London; Dr David J. Drewry of the Scott Polar Research Institute; Dr John A. Gulland from the Food and Agricultural Organization (FAO) and Imperial College. The views of diplomats and lawyers who have been actively engaged in the development of the Treaty System in recent years were there: Mr John Rowland (Australia); Mr Rolph Trolle-Andersen (Norway); Dr John Heap (UK). The scientific and government viewpoints are not the only ones to be included. The question of commercial development was discussed by Mr F. G. Larminie, Environmental Control Centre, British Petroleum (BP). Mr Jim Barnes, Director, The Antarctic Project, Washington DC, strongly puts the case for the environmentalists and H. E. Zain-Azraai, the

Permanent Representative of Malaysia to the United Nations (UN), speaks of the Third World and explains the thinking behind the recent initiative in the UN to increase its involvement in Antarctica. Highly qualified chairmen presided over the sessions and included Sir Vivian Fuchs (formerly Director of the British Antarctic Survey), and Dr Tore Gjelsvik (formerly Director of the Norwegian Polar Research Institute).

Chairman of the whole conference and the closing session was Sir Ian Sinclair who used his wide knowledge and experience gained as Legal Adviser to the FCO to steer the discussions.

Lord Shackleton   23 January 1986

Introduction

Antarctica, known to most people as an ice blue and beautiful but isolated and fragile continent, has long been of interest to scientists, explorers and a handful of diplomats. Its history is redolent with dramatic images of courage and comradeship, tragedy and triumph. The exploitation of marine living resources, and more recently mineral and oil resources have, however, been the dominant forces behind international interest in Antarctica. Enticed by Cook's voyages, seal hunters explored the South Antarctic islands in the late eighteenth century. Over-exploitation of seal herds lead others in the 1830s to move closer to the Antarctic continent for elephant seals and whales. Today, the depletion of fish stocks and the expansion of coastal state sovereignty over 200 mile Exclusive Economic Zones has again focussed attention upon the southern oceans, this time to harvest krill. Extravagant hopes for mineral and oil wealth, inflamed by negotiations for an Antarctic minerals regime, have directed international attention to Antarctic non-living resources, thereby prompting fears for the conservation of this fragile and beautiful continent.

It was at a time when commercial interests in Antarctica had declined, and when individual scientific research was highly esteemed, that 12 states were able to negotiate the Antarctic Treaty in 1959. Compelled by the need to avert tension and disorder and by a determination that scientific research should continue unimpeded, these states successfully avoided the formidable legal issue of sovereignty and created the Antarctic Treaty regime of interlinked conventions and recommendations. This regime is justifiably hailed as a remarkably effective international system which has largely achieved its relatively modest objectives. Times, however, have changed. Once again international attention has turned to Antarctica as a resource-rich continent, for Antarctica is a valuable source of krill, and other marine resources, and expectations have been raised for its offshore oil and gas potential.

Further, and opposed, concerns are now expressed for the preservation of the unique Antarctic environment once resource exploitation, particularly of non-living resources, proceeds. The voyage of the *Greenpeace* vessel to Antarctica in 1985–6, whatever the dangers it courted, has served to dramatise fears which have been generated by negotiation of a minerals regime and to publicise demands that Antarctica be declared a world park or wilderness zone.

The United Nations has included the 'question of Antarctica' on its agenda and is now concerned to develop policies for the future regulation of Antarctica. Encouraged by the success of negotiations to achieve the 1982 Convention on the Law of the Sea, some states in the General Assembly have argued that Antarctica is the common heritage of mankind and should be subject to international regulation so that the benefits of resource development accrue to all states on an equitable basis. The General Assembly has considered the question of Antarctica at its 38th, 39th and 40th sessions, and on 27 November 1985 the General Assembly requested the Secretary-General to submit his expanded study on Antarctica to the 41st session in 1986. In particular, it was resolved that the Secretary-General should report on the Antarctic Treaty Consultative Parties negotiations on a minerals regime on the ground that 'any exploitation of the resources of Antarctica should ensure... the international management and equitable sharing of the benefits of such exploitation'.

When introducing the collection of papers of which he was editor, *Antarctic Resources Policy: Scientific, Legal and Political Issues* (CUP, 1983), Francisco Orrego Vicuña pointed out that science, economics, law and politics have begun to converge in relation to Antarctica at a 'dizzying speed'. It is this dizzying speed which explains why, 3 years later, a further collection of papers concerning Antarctica warrants publication.

The papers which are the subject of the present volume were written for an international conference held in London, 11–12 April, 1985. The conference, entitled 'Whither Antarctica?', was organised by the British Institute of International and Comparative Law under the directorship of Lady Fox. This work is intended to build upon the Orrego publication by updating the legal, resource and environmental issues presently under consideration within the Antarctic Treaty system and, more recently, by the Secretary-General of the United Nations. An objective of this collection has been to provide a guide to the papers by including an introduction to each part which incorporates points made during conference discussion. The intention is that the reader should have some sense of the diversity of opinion on each topic. This work is prepared under the General Editorship of the British Institute of International and Comparative Law.

This publication owes much to the experience and assistance of Dr B. Stonehouse of the Scott Polar Research Institute for which the editor thanks him.

Part I: Antarctica: physical environment and scientific research

The first part includes two papers which consider the unique physical environment in Antarctica and the scientific research which is and can be undertaken there. Dr D. J. Drewry, Director of the Scott Polar Research Institute, describes the geographic and geological characteristics of Antarctica and emphasises the special influences of the continental ice cover upon the earth's climate, oceanographic patterns and the unusual depth of the Antarctic continental shelf. Dr R. M. Laws of the British Antarctic Survey outlines the diversity and interrelated nature of scientific research in Antarctica which influences the study of geology, geophysics, plate tectonics, glaciology, climatology, oceanography, meteorology and geophysics, biology and ecology of living organisms. He argues that the very special nature of Antarctic research lies in the simplicity of the environment. There are few people, no industry; environmental impact and pollution are minimal; the rock structures are relatively uncomplicated; the ecosystems are non-specific and the ocean food webs are simple and dominated by krill as the key species. This simplicity, Laws believes, provides scientists with unique opportunities to expand their knowledge.

Part II: The Antarctic Treaty regime: legal issues

The second part considers the Antarctic Treaty system and its interlinked Recommendations and Conventions, including the Convention on the Conservation of Antarctic Marine Living Resources (CCAMLR). Mr Rolph Trolle-Andersen, of Norway's Ministry of Foreign Affairs, describes the negotiations leading to the Antarctic Treaty in 1959. He explains the Treaty's salient features, most particularly the 'ingenious formula' of Article IV, which preserves the legal positions of the Treaty Parties on the question of territorial sovereignty. He describes the practice of consensus in Consultative Party meetings, the 'legislative' procedure of Recommendations and the separate negotiations of inter-linked Conventions. In response to criticism that Antarctica is a secretive club of states, Trolle-Andersen stresses the cooperative and evolutionary nature of the Antarctic Treaty system which, he argues would be impossible to recreate in today's political climate. A similar theme is taken up by Professor Francisco Orrego Vicuña, Director of the Institute of International Studies, University of Chile, who considers the Antarctic Treaty system is the only viable alternative to regimes based on claims to territorial sovereignty or opposing assertions of universal management. He emphasises the advantages of regulation through the Antarctic Treaty system in contrast with wider international participation and counters criticism of the system, including those concerning the decision-making processes and lack of access to information. The question of the right of the Consultative Parties to regulate activities in Antarctica is met with a further question, on what basis does the international community, through the United Nations General Assembly, assert a superior right?

Both Orrego Vicuña and Trolle-Andersen raise questions of concern to the international lawyer though, for Antarctic scientists and explorers, the intervention of lawyers in Antarctic matters has been viewed with scepticism and some hostility. Lady Fox, Director of the British Institute of International and Comparative Law, asks what contributions, if any, a lawyer can make to Antarctic regulation. She nominates four stages at which legal skills have value. The first concerns the validity of claims to territorial sovereignty which are advanced in accord with classical theories of acquisition. The second concerns the negotiation of techniques to accommodate the diverse legal interests of claimant, non-claimant and non-recognising States. The third relates to development of the legislative processes of regulation through the Antarctic Treaty system. The fourth, and most intellectually challenging stage, involves the fashioning of new concepts of territorial jurisdiction to resolve or diffuse these conflicting demands. Her paper draws attention to the interests which should be considered by an international lawyer when assessing models for Antarctic-resource management, including commercial, conservation, recreational and strategic interests which remain outside the ambit of current Antarctic regulatory processes.

The final paper in this part, Dr Gillian Triggs, takes up some of the legal questions raised by Fox, most particularly those jurisdictional problems which owe their origin to the fundamental conflict between states over territorial sovereignty. These are first, jurisdictional ambiguities which arise from the relationship between the 1982 Convention on the Law of the Sea and the Antarctic Treaty system; second, the criticisms of limited access to decision-making within the Antarctic regime; and third, the legal implications for Antarctica of the concept of a common heritage of mankind. This last aspect provides a legal background to the subsequent discussion by Ambassador Zain-Azraai who describes a Third World approach to the concept of a common heritage.

Part III: The Antarctic Treaty regime: protecting the marine environment

The third part is concerned with protection of the marine environment. It is introduced by a summary of the Convention on the Conservation of Antarctic Marine Living Resources (CCAMLR). Dr John A. Gulland, of the United Nation's Food and Agricultural Organization (FAO) and the Imperial College, London, outlines the interlinked food chains and life cycles of Antarctic marine species within the Antarctic Convergence. He describes some of the lessons which have been learned in past attempts to conserve and manage marine resources, particularly by the International Whaling Commission. He traces the influence which the Antarctic Treaty has had upon the structure of the Conventions on Sealing and Antarctic Marine Living Resources and concedes that measures adopted under CCAMLR to conserve Antarctic fisheries may appear 'cosmetic'. He also takes the position that, imperfect though regulation may be under the Antarctic Treaty system, it remains the preferable alternative to universalisation.

Dr Martin Holdgate, Chief Scientist at the Department of the Environment, London, develops the thesis that the exploitation of living resources must be regulated. He documents the failure to control Antarctic fishing in the past, a failure which derived from unregulated exploitation of a common property of 'open-access' resource. He describes existing environmental regulation in Antarctica as establishing both general principles of conservation and use and a specific legislative process for recommendations and measures in certain areas of human impact. Holdgate concludes that the Antarctic Treaty system has evolved methods of environmental conservation which, at the same time, facilitate scientific research. While he lists defects in the system of environmental protection, he concludes that these defects can be resolved by change within the system itself. Mr William N. Bonner, Head, Life Science Division at the British Antarctic Survey, provides a history of Antarctic Treaty Parties' attempts to regulate the environment and describes the present negotiations between the Scientific Committee on Antarctic Research and the International Union for the Conservation of Nature (IUCN) on World Conservation Strategy. He makes the dramatic point that the Antarctic is already, by far, the largest natural reserve in the world.

Mr James N. Barnes, Director, Antarctic Project, Washington, DC, has

contributed a challenge to those authors who call for reform within the
Antarctic Treaty system, by arguing that Antarctic resource policies must
be developed in light of evolving interests within the international community. He describes the roles of non-governmental organisations, such as
the Antarctic Southern Ocean Coalition, in the formulation of resources
policy, and commends the formation of an Antarctic Environment
Protection Agency. He further argues for a total moratorium on all
minerals activities. The paper is critical of existing regulations on Antarctic
conservation, most particularly of the proposed construction of an airfield
in Adelie Land and of the limited scope and number of protected sites
within Antarctica.

Part IV: The Antarctic Treaty regime: minerals regulation

The current negotiations on a minerals regime by the Antarctic
Treaty Consultative Parties provided the 'leitmotiv' of the conference in
the sense that proposals for a legal regime to regulate the exploitation of
minerals concentrate the mind upon all aspects of Antarctic management
and provide a focus for the development of the concept of a common
heritage of mankind. The first paper by Mr F. G. Larminie of the
Environmental Control Centre, British Petroleum, provides a salutary
caution to the grandiose expectations of those proclaiming a minerals
boom in Antarctica. He argues that prospects for Antarctic minerals
exploitation could well be considered 'virtually nil'. He argues that the
negotiation of a minerals regime has no commercial motive and does not
reflect pressure from the petroleum industry. He warns against the
negotiation of an overly detailed convention which cannot deal with, as
yet, hypothetical problems. Rather, he urges a general principles, evolutionary approach to the establishment of an agreement. Mr Arthur Watts,
Legal Counsellor, Foreign and Commonwealth Office, London, also
warns that there is little likelihood of commercial exploitation of mineral
resources in Antarctica. He explains that negotiations have been undertaken in order to establish a regime well in advance of technical developments and economic viability. He notes (the irony) that negotiations are
easier when so little is known about mineral deposits and commercial
prospects appear non-existent. Watts emphasises that, in principle, mineral
activities will be prohibited under a negotiated regime unless stringent
procedural and substantive requirements are met. He then provides an
outline of the structure of a minerals regime which already attracts a
certain measure of consensus.

Dr Gillian Triggs then considers a negotiating draft for a minerals
regime which is now publicly available. While this draft does not represent

a final document, it does suggest the kinds of mechanisms which are under consideration for resolution of conflicting interests concerning sovereignty, conservation of the environment and third world access to resources.

Part V: Whither Antarctica? Future policies

Dr John A. Heap considers the question of future problems within the Antarctic Treaty system and does so in the light of the preambular objective of the Antarctic Treaty that Antarctica should not become the 'scene or object of international discord'. He argues that fears that the Antarctic Treaty Parties will be unable to deal with allegations of infringements of Treaty obligations (for example, of French plans to build an airstrip at Pointe Geologie) raise further doubts whether Parties can properly deal with prima-facie infringements of rules established within a future minerals regime. However, he believes that a more rigorous inspection system and the development of the political will to favour long-term interests above short-term advantages can overcome these problems. His paper considers the specific difficulties raised by fishing, minerals exploitation and decision-making (which rests upon the lowest common denominator) and claims to territorial sovereignty. He concludes, as do earlier writers, that effective management of resources and the environment can be achieved if states assert their claims to territorial jurisdiction rather than through any form of internationalisation.

It is this conclusion which concerns Ambassador Zain-Azraai who argues that Antarctica is, or should become, the common heritage of mankind. His paper criticises the assumption that a right to regulate Antarctic resources can be founded on prior experience. Rather, his paper argues that the justifiable interests of the international community in Antarctic management provide an appropriate basis for assuming a right to regulate resource exploitation in Antarctica in the future. Zain-Azraai recommends the formation of a United Nations Ad Hoc Committee to consider future regulation of Antarctic resources on the ground that to do so provides the best chance of reconciling all legitimate interests.

The opposing views of Ambassador Zain-Azraai and Dr Heap are considered by Mr J. R. Rowland in his paper which assesses alternative strategies for Antarctic regulation. He considers proposed reforms within the Treaty system itself, and more radical suggestions for a United Nations Trusteeship for a new international regime negotiated under the auspices of the United Nations and which has a genuinely universal character, and for separate regulation of minerals exploitation by a new international body representing all special interests. Rowland observes that the United Nations General Assembly may well endorse the central articles of the

Antarctic Treaty and possibly also Article IV. Nonetheless, he fears risks of instability which are posed by negotiations for a minerals regime. He favours a moratorium on exploitation and the declaration of Antarctica as a world conservation area subject to United Nations endorsement. This, he argues, is a practical means of avoiding conflicts between advocates of a common heritage, conservationists and Antarctic Treaty Parties.

The United Nations in Antarctica? A watching brief

In conclusion, the editor draws the papers together by summarising the legal, environmental and political issues confronting the Antarctic Treaty parties, and identifies those aspects of the Antarctic Treaty regime which require change or modification. A summary is given of the views of non-Treaty states which were submitted to the Secretary-General and included in his report to the United Nations in 1984. Consideration is given to the procedural options open to the General Assembly when it reviews the 'Question of Antarctica' on the agenda in 1986. While it seems unlikely that any Antarctic committee will be established under the auspices of the United Nations, the General Assembly may adopt a 'watching brief' over activities within the Antarctic Treaty regime by retaining the issue on the agenda.

Gillian D. Triggs

The contributions published in this book were prepared for the Conference on Antarctica, 'Whither Antarctica?', organised by the British Institute of International and Comparative Law, 11–12 April 1985, in London. The Conference was assisted by a generous contribution from Britoil p.l.c. under the Chairmanship of Sir Phillip Shelbourne for which the Institute is most grateful. A follow-up meeting was held on 4 March 1986, again at the British Institute of International and Comparative Law.

Part I
Antarctica: physical
environment and scientific research

1

Introduction

Antarctica is the coldest, windiest and most inhospitable continent in the world, which differs from the Arctic in being a land-based continent and in having no indigenous population. Dr Phillip Law describes Antarctic activities as occurring during four periods: the trading, imperialist, scientific and resource eras ((1985) **10** (4) *Interdisciplinary Science Review*, 336–8). International interest in Antarctica began with commercial whaling and sealing expeditions through the eighteenth and nineteenth centuries. This trading era was followed by an imperialist era from 1890 to the 1940s which was a period of high adventure, heroism and colonial territorial aggrandisement, though predominately of a marine nature. Expeditions during this period were amateur and were engaged in preliminary scientific work collecting, describing and classifying information. It was not until the International Geophysical Year (IGY) of 1957–8 that scientific interests were generated in earnest. During the IGY, 50 stations were set up over the Antarctic continent and comprehensive scientific work was undertaken. Research during the IGY was subsequently maintained through a permanent committee of the International Council of Scientific Unions (ICSU), the Special Committee on Antarctic Research (SCAR). One of the dominating forces for the negotiation of the Antarctic Treaty in 1959 was to ensure that territorial claims should not retard comprehensive scientific research. With this objective, the Parties agreed that:

> freedom of scientific investigation in Antarctica and co-operation toward that end, as applied during the International Geophysical Year, shall continue.

It was further agreed that all information regarding scientific programmes in Antarctica should be exchanged, that scientific personnel should be exchanged between expeditions and stations and that scientific observations and results should be exchanged and made freely available. These undertakings have become cornerstone principles of the Antarctic Treaty system and those who have worked on Antarctic bases testify to the unique cooperation demonstrated there between scientists of all nationalities.

While international and governmental interest is now concerned with problems of Antarctic-resource development and regulation, it remains imperative that scientific research activities be maintained and increased; for, despite the quality and extent of research over the past 30 years, there continue to be significant gaps in our knowledge about Antarctica. As a natural resource, Antarctica presents scientific challenges as great as those presented by the oceans, the atmosphere, space and the moon and other celestial objects. Indeed, a clear understanding of the Antarctic is a necessary part of our understanding of the globe's environmental system.

The following chapters describe the physical environment of the Antarctic continent and surrounding waters and the research which is currently being conducted there. They stimulated discussion amongst participants at the British Institute's Conference about the future direction of scientific research as a means of facilitating exploration and exploitation for marine and non-living mineral resources. It was accepted that there was no readily discernable distinction between scientific-resource exploitation and pure scientific research. The difficulties of distinguishing between the two were seen as posing a potential conflict between the Treaty-protected right to free scientific investigation and the need to regulate, and possibly to prohibit, scientific research for exploitation purposes. This apparently intractable problem arose in discussion later in the context of negotiations for a minerals regime.

The high cost of research was noted, much of the budget for some states being spent in transporting scientists each season to and from Antarctica. There was also a tendency for new states to locate their bases alongside existing ones. The requirement in Article IX that, in order to gain consultative status, acceding states must conduct substantial scientific research in Antarctica was an obstacle to the participation in the Treaty regime of small states with limited resources. Indeed, some of the original 12 Consultative States have found recent programmes costly as events in the 1985/6 season show with the sinking of the *Southern Quest* and besetment of the *Nella Dan*. Although a desirable feature of the present system has been the willingness of states to make a financial outlay on scientific research with no immediate commercial return, increased cooperation and shared spending of large scientific projects were indicated in the future to spread the cost and to ensure participation by more states in serious scientific research in the Antarctic.

It was also noted in discussion that the logistical problems of Antarctic research led to a high degree of involvement of government agencies, particularly of the armed forces. This involvement raises fears as to the use of military personnel and equipment, in the light of Article I that Antarctica shall be used for peaceful purposes only, and that the establishment of military bases and fortifications, or the carrying out of military manoeuvres are prohibited.

2

The Antarctic physical environment

D.J.DREWRY

Introduction

Antarctica forms a unique yet integral part of the environmental systems of planet earth. Located asymetrically around the south geographical pole it constitutes an area of 13.918 M km², almost twice the size of

Fig. 2.1. General location map for Antarctica indicating the principal elements referred to in the text. This map should be used in conjunction with Fig. 2.8 which depicts Antarctica in relation to the Southern Ocean and parts of the other southern hemisphere continents.

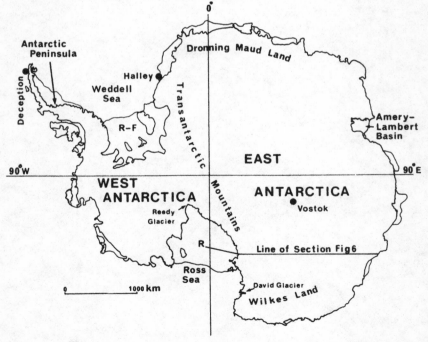

Australia and 57 times the size of the United Kingdom. (Fig. 2.1). Only
2.5% of Antarctica is ice-free, the remainder being covered by a permanent
ice mass. During the austral winter, sea ice rapidly develops around the
continent creating an additional area of ice-covered ocean of some
20 M km². This extensive ice cover is probably the single most distinctive
feature of Antarctica. It makes the continent an ideal laboratory for
studying the influence of ice in modulating global climate, and in modifying
oceanographic processes. The investigation of the behaviour of the
land-based ice sheet assists in assessing the general role of ice in the
geological shaping of our planet at present and in the past.

Geographical setting

Geographically there are two distinctive definitions of Antarctica.
First, and most obvious, is that based upon the existing ice sheet which
extends from sea level to just over 4000 m in the centre of East Antarctica.
Beneath this relatively smooth ice carapace, however, lies the second, more
permanent Antarctica – the continental bedrock. A variety of remote
sensing techniques have, during the last three decades, assisted the
mapping and investigation of these two terrains which are depicted in
Fig. 2.2, as currently understood.

It is apparent that the subglacial bedrock surface reveals a diversity
which is comparable with any other continental area, possessing major
mountain chains and extensive basins. This topography reflects, in part the
geological composition and deeper structure of the Antarctic crust. East
Antarctica forms a land mass, largely above sea level with a number of
mountain ranges. The Gamburtsev Mountains are the largest within the
continental interior and rise to elevations of 3800 m. Many other smaller
blocks and massifs are known, although exploration of Dronning Maud
Land is still fragmentary. The mainly exposed Transantarctic Mountains
form a major topographic high along the Pacific edge of the East
Antarctica plate extending some 3500 km from Cape Adare to isolated
ranges close to the Filchner Ice Shelf. Basins are principally located in
Wilkes Land – the Wilkes and Aurora subglacial basins – extending in
places to well over 1000 m below sea level. Smaller troughs are common-
place, some with spectacular flanking escarpments. Much of West Antarctica, in contrast, lies well below sea level. In the extensive Byrd
subglacial basin elevations reach to 2500 m below sea level. Mountain massifs are found along the Pacific coast, in the rugged Antarctica Peninsula
and in a block running inland (west) of the exposed Ellsworth Mountains,
where the continent's highest peak is located (Vinson Massif, 5140 m).

The present-day configuration of the continent described above is a

Fig. 2.2.(a) Isometric view of the surface of the Antarctic ice sheet viewed from the longitude of 140 °E. The grid is of cell size 50 km and is fitted to all known surface elevation measurements (see Drewry, 1983). Note the pronounced topographic effect of ice drainage into the Lambert Glacier–Amery Ice Shelf system, the Ross and Weddell Sea embayments and the sinuous spine of Antarctic Peninsula. Major mountain chains are not depicted. The grid datum is sea level. (b) Isometric view of the subglacial bedrock surface of Antarctica. The view point is the same as in Fig. 2(a). The topography has been considerably smoothed which has resulted in only major features being portrayed. Note the major mountain ranges in central East Antarctica (Gamburtsev Mountain), the substantial areas below sea level in West Antarctica (Byrd Basin) and in Wilkes Land. The grid datum is sea level.

(a)

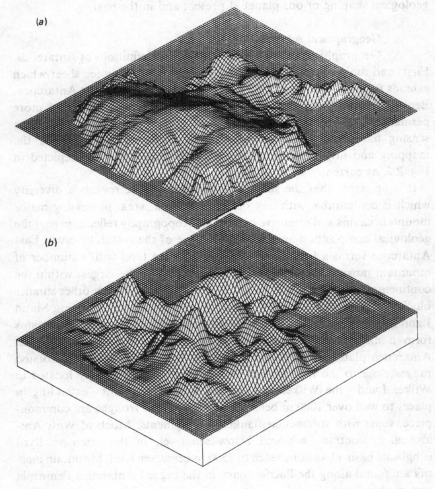

(b)

distorted version of the 'true' bedrock surface due to the weight of the overlying ice sheet which depresses the relatively rigid lithosphere into the denser and more viscous asthenosphere by an amount roughly proportional to the ratio of the density of the ice to that of the mantle (approximately 1:3). Isostatic readjustment by computer allows the 'real' surface of Antarctica to be established. Up to 1000 m of uplift occur in the central areas of East Antarctica and 500 m in central West Antarctica. Although the pattern of readjustment is complex due to the rigidity of the crust and significant variations in loading, some of the major topographic relationships remain unchanged. The locations of the new 'sea level', and, hence, the land areas above and below adjusted sea level, undergo significant alteration.

Beyond present-day sea level, the continental shelf extends around much of East Antarctica as a relatively narrow feature of a few hundred kilometres width (Figs 2.4 and 2.8). The only areas of significant development are in the Ross and Weddell Sea embayments and the little investigated regions of the Bellingshausen and Amundsen Seas off West Antarctica. By comparison with many continental shelves, that of Antarctica is particularly deep, on average at about 500 m below sea level. Such depths are probably attributable to sediment starvation during the last 20 Ma or so when ice sheets on land inhibited discharge of sediments from the continent. The depth, complex morphology and low sediment thickness have implications, mostly adverse, for the future prospectiveness of the Antarctic continental shelf for hydrocarbons.

The edge of the continental shelf leads down into the abyssal plains of the surrounding ocean areas at depths of over 4000 m below sea level. The seafloor is veneered with sediments, thinning towards the high mid-ocean ridge where new, volcanic ocean floor is constantly being created. The oldest parts of the circum-Antarctic ocean relate to the earliest period of rifting of the former Gondwana supercontinent, and are of the order of 150 Ma old. The mid-ocean ridge is offset by numerous transform faults which allow the complex differential motions of the continental plates during ocean-floor spreading to be accommodated. Unlike most other major plates, the Antarctic is almost totally surrounded by passive margins; only in the Scotia Sea region is there significant compressive motion and subduction.

Geophysical structure

Seismic surface wave dispersion studies, deep seismic refraction investigations and the interpretation of gravity observations have demonstrated the continental nature of Antarctica. They also reveal considerable crustal complexity (Bentley, 1973; 1983; Yukutake & Ito, 1984). Distinctive differences are apparent in the thickness of the crust (defined as the depth to the Mohorovicic discontinuity where there is a significant increase in seismic compressional, 'P'-wave, velocity) as shown by the results of gravity calculations. Although different studies diverge in their results it is clear that East Antarctica possesses a crustal thickness which is an average 30–40 km, reaching greater values in the region of major mountain chains such as the Gamburtsev Mountains, whilst West Antarctica is some 10–15 km thinner. The boundary between these two 'provinces' lies along the present trace of the Transantarctic Mountains.

The detailed structure of the upper crustal layers has only been investigated with any reliability in a few places. Soviet geophysicists have undertaken deep refraction seismic work in the vicinity of Novolazarevskaya and in the Amery Ice Shelf–Lambert Glacier region (Fig. 2.3) (Kogan, 1972; Kurinin & Krikurov, 1982) and Japanese scientists in the Mizuho Plateau area (Ikami et al., 1983). Interpretation of the velocity profiles suggests a normal crust with an upper layer of some 20 km thickness of relatively high velocity indicative of old Pre-Cambrian crystalline basement.

In the Amery–Lambert Rift zone a graben extends inland several hundred kilometres and in which the crustal thickness is only 25 km, with a 5 km uppermost section of sediments. Shallower seismic sections elsewhere in the continent yield some general knowledge concerning the composition of the upper few kilometres of the crust. Combined with magnetic measurements they are able to reveal, for instance, regions dominated by crystalline basement rocks and the presence of major sedimentary basins lying above the basement (Drewry, 1978; Jankowski et al., 1983). Such latter features are found both in East Antarctica and in the Ross and Weddell Sea embayments (Davey et al., 1983).

Geological evolution and mineral resources

With an area of rock outcrop less than 2.5%, it has proved a major challenge to geologists to piece together the geological history of Antarctica. Given the fragmentary nature of the evidence, in both space and geological time, represented by isolated nunataks, small coastal exposures and often precipitous mountain regions, it is remarkable that a relatively comprehensive picture has emerged in recent years. This picture is the more astonishing in view of the hostile Antarctic environment.

Antarctica is formed of an ancient block or shield of Pre-Cambrian to early Palaeozoic strata which has been metamorphosed and extensively

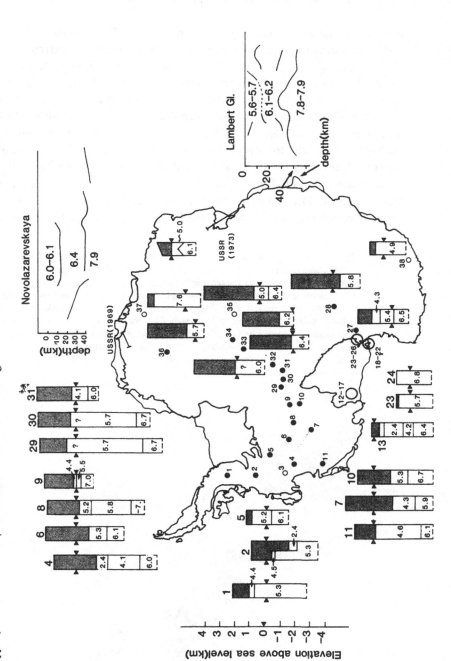

Fig. 2.3. Crustal structure of Antarctica from seismic refraction profiling (after Yukutaka & Ito, 1984). The columns show the results of refraction seismic shooting with velocity interpretations in the rock layers beneath the ice (ice is shown stippled). Velocities are in km/s. The locations of the columns are given by dots and numbers. Two deep refraction seismic sections are shown for an area near Novolazarevskaya and the Lambert–Amery Rift undertaken by Soviet scientists. Numbers again indicate 'P'-wave velocities.

intruded by igneous (mainly granitic) rocks (Fig. 2.4) (James & Tingey, 1983). Radiometric dating suggests some of these rocks are extremely old in the order of several thousand million years. It is highly probable that the shield, which possesses an areá in the order of 9.9 M km², is composed of a number of fundamental stable units or cratons in a similar manner to the Pre-Cambrian shields of Western Australia, Southern Africa, India and South America. Detailed differentiation of the shelf which embodies perhaps 85% of the total geological time span of Antarctica is seriously hampered, however, by being almost entirely covered by the ice sheet.

The Pacific margin of East Antarctica is characterised by more complete exposure in the Transantarctic Mountains which has revealed evidence of successive orogenic activity from mid Pre-Cambrian to early Palaeozoic

Fig. 2.4. Schematic tectonic map of Antarctica based, in part, upon Craddock (1972), and Cameron (1978). Key: 1, East Antarctic shield; 2, late Pre-Cambrian and Early Palaeozoic orogenic zones of Transantarctic Mountains (Beardmore and Ross); 3, Mid-Palaeozoic orogenic zone of Northern Victoria Land (Borchgrevink); 4, early Mesozoic orogenic zone of Ellsworth Mountains (Gondwanian); 5, late Mesozoic to early Cenozoic orogenic zone of Antarctic Peninsula and Byrd Land (Andean); 6, subglacial sedimentary basins of Ross Sea, Byrd Basin and Weddell Sea; 7, East Antarctic intracratonic sedimentary basins; 8, Cenozoic volcanics. The dotted line is the approximate edge of the Continental Shelf.

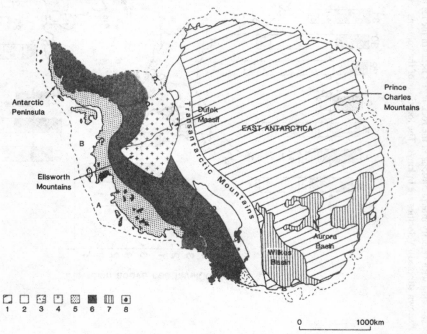

times. The most recent widespread episode is termed the Ross orogen of 400–500 Ma age. Its structural style and age are comparable with events in the Adelaide geosyncline of South Australia. Upon the eroded and uplifted terrain of the Ross orogen a thick sequence of terrigenous sediments was deposited commencing in early Palaeozoic time extending over land mass which comprised India, South America, Australia, Antarctica, New Zealand and Malagasy, and called Gondwana (Fig. 2.5). Little deformed, these rocks are termed the Beacon Supergroup in Antarctica and, with similar units in the Gondwana continents they demonstrate a common history until Mid-Mesozoic times when the supercontinent commenced to rift and break into its now familiar continental fragments.

Disruption was accompanied by major igneous activity. The timing of this complex continental separation is revealed in the magnetic anomaly patterns of the ocean floors around Antarctica. A distinctive deformation period affected the Northern Victoria Land region in Devonian to early Carboniferous times and has affinities with orogenic activity in Tasmania and eastern Australia. In the Ellsworth Mountains of West Antarctica deposition of the thick sequence of rocks similar to the lower part of the Beacon Supergroup was terminated by orogenic activity in late Triassic to early Jurassic time. The deformation has strong similarities with that of the Cape Fold Belt of southern Africa. Against the periodically deforming proto-'Pacific' margin of Antarctica further geosynclinal belts formed during Mesozoic time and were subject to orogenic activity in late

Fig. 2.5. Principal tectonic elements of Gondwana. The continental reconstruction is based upon Gondwana Five (1981). Main pan-Gondwana zones are indicated although their boundaries and areas remain uncertain in Antarctica.

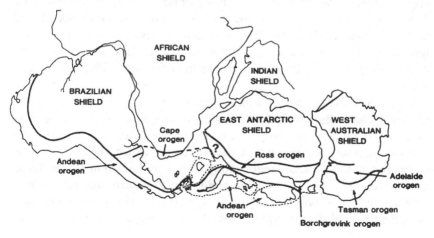

Mesozoic–early Cenozoic times with widespread intrusive activity, but little metamorphism. This 'Andean' phase is coeval with the widespread orogenic zones of South America (Dalziel, 1982; Thomson et al., 1983).

In Cenozoic times the latest chapter of Antarctica geological history has been characterised by the occurrence, principally in West Antarctica, of abundant basaltic volcanism. By the end of the Mesozoic (65 Ma BP) much of the present configuration of Antarctica and its near polar position had been established. The opening of the Southern Ocean by seafloor spreading led to the progressive isolation of Antarctica from the other continents and development, once the Drake Passage was created, of the circum-Antarctic current. This, more than any other factor, was probably responsible for progressive refrigeration of the continent and growth of the ice sheet in Mid-Cenozoic times (Kennett, 1979).

Mineral resources

Knowledge of Antarctic earth history allows some brief comments to be made on potential mineral resources. The Antarctic, at times possessing a common history of geological development with other southern hemisphere land masses is likely to possess (at least on statistical grounds) similar assemblages of minerals. Several studies have been made, based upon comparisons of mineral occurrences within specified geological provinces, between Antarctica and the other Gondwana fragments (Wright & Williams, 1974). Other assessments are formulated in terms of actual mineral occurrences. A full range of metalliferous and non-metalliferous ones are known and are summarised by Rowley et al. (1983) and de Wit (1985).

Particular attention may be placed upon the Antarctic Peninsula which appears heavily mineralised in a complementary manner to the southern Andes. Another region of possible mineral potential is the Dufek Massif of the Pensacola Mountain. This mountain block is a Jurassic-layered igneous intrusion which recent geophysical work has shown to be of comparable size to the Bushveld complex in South Africa (Behrendt et al., 1980).

Hydrocarbons are typically found in relatively young sedimentary rocks in basins whose characteristics have allowed adequate maturation, migration and trapping. Comments with regard to the nature of the Antarctic continental shelf suggest only a limited potential for these conditions (Cameron, 1981). Onshore sedimentary basins are known in Wilkes Land and are comparable with similar regions in Southern Australia which produce hydrocarbons. The thick ice cover (up to 4 km), however, must limit their possible future exploitation. Permian coal is widely known from the Transantarctic Mountains and the Prince Charles Mountains. Seams are often thin and discontinuous and the coal is of low rank having been subject to heating by Jurassic dolerite intrusions (Whitby, Rose & McElroy, 1983). Their potential, however, cannot be overlooked.

Ice sheet

Antarctica is the most glaciated of all continents. It contains 30.11 M km³ of ice distributed in the large interconnected ice sheets in East Antarctica, West Antarctica, the flanking floating ice shelves in the Ross and Weddell Seas and the Antarctica Peninsula (a region of complex glacierisation; Drewry, 1983b; Fig. 2.2a). The distribution of this ice is given in Table 2.1.

Ice is probably the most individual characteristic of Antarctica. It dominates the present-day environment exercising fundamental control on climate and ocean process and hence of biological activity. It has already been demonstrated to exert a significant stress on the lithosphere and depress it by upwards of 1000 m.

Figure 2.6 provides a typical view of the ice sheet and its key elements. The ice is fed by snow accumulation which varies from a few centimetres per year in central regions far removed from oceanic moisture sources to several tens of centimetres to a metre in near-coastal localities. The interior of the Antarctic is thus, a cold desert. Under normal circumstances the ice,

Fig. 2.6. Section through the Antarctic Ice Sheet (see Fig. 2.1 for location) depicting the East Antarctic Ice Sheet and part of the Ross Ice Shelf. Note the rugged subglacial topography. To the right, snowfall, paths traced by particles of ice within the ice sheet and typical surface velocities are shown schematically. Note the increase in speed from centre to periphery (partly after Drewry, 1983).

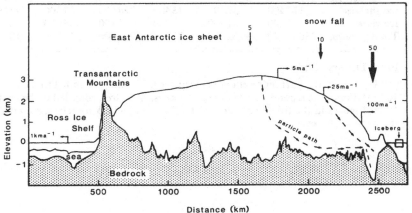

Table 2.1. *Areas, ice thickness and ice volumes for the principal regions of Antarctica*[a]

	East Antarctica	West Antarctica	Antarctic Peninsula	Ross Ice Shelf	Ronne–Filchner Ice Shelves	Total
Grounded ice	9855570	1809760	300380			11965700
Ice shelves	293510	104860	144750	525840	472760	1541710
Ice rises	4090	3550	1570	10230	59440	78970
Total ice	10153170	1918170	446690	536070	532200	13586400
Rock outcrop	200630	55970	75090			331690
Total area	10353800	1974140	521780	536070	532200	13918070

	East Antarctica	West Antarctica	Antarctic Peninsula	Ross Ice Shelf	Ronne–Filchner Ice Shelves	Mean
Grounded ice	2630	1780	610			2450
Ice shelves	400	375	300	427	650	475
Ice rises	400	375	300	500	750	670
Mean	2565	1700	510	430	660	2160

	East Antarctica	West Antarctica	Antarctic Peninsula	Ross Ice Shelf	Ronne–Filchner Ice Shelves	Total
Grounded ice	25920100	3221400	183200			29324700
Ice shelves	117400	39300	43400	224500	307300	731900
Ice rises	1600	1300	500	5100	44600	53100
Total volume	26039200	3262000	227100	229600	351900	30109800

From Drewry, 1983*b*.

[a] Figures in first part represent area in square kilometres, rounded to the nearest 10 km². Figures in second part represent thickness in metres, rounded to the nearest 10 m. Figures in the third part represent volume in cubic kilometres, rounded to the nearest 100 km³.

behaving as a viscous-plastic material, flows outwards from the continental interior under the influence of gravity. Its speed is so adjusted to be just sufficient to balance and evacuate the incoming snowfall (when averaged over many centuries). In this way the ice sheet is able to maintain an overall quasi-parabolic profile. The central regions are consequently very flat with gradients in the order of 10^{-3} and where the flow is sheet-like (Fig. 2.2a). Towards the ice margin velocities are greater, the ice is thinner and there is significant transmission of stresses from flow over the irregular subglacial bedrock. Here the ice surface is rough and undulating. Within 200–300 km of the coast, the ice may become channelled, either through peripheral mountains (as in the Transantarctic Mountains) where outlet glaciers develop (Fig. 2.7) or as fast-flowing zones within the ice sheet itself and

Fig. 2.7. Typical outlet glacier (Reedy Glacier) of the Transantarctic Mountains (See Fig. 2.1 for location). The ice sheet is in the distance; the ice flow is towards the foreground. Note surface flowlines.

termed ice streams. Such latter features are found around much of the continental periphery but are best developed in Marie Byrd Land (McIntyre, 1985).

In some instances, the ice discharging from the continent flows into the sea. Here it comes afloat and continues to move outwards under its own weight and that of surface snowfall in the form of ice tongues (Fig. 2.7) and ice shelves. The inner parts of the Ross and Weddell Sea embayments are filled with such ice shelves which attenuate in thickness from close to 1000 m near the grounding line (the point where the ice comes afloat) to about 200 m or less at the ice front. At the bottom of ice shelves the ice is in direct contact with sea water. Depending upon the temperature contrast between water and the ice, the ice shelf base may melt and thin, or thicken by the accretion of small ice crystals to its underside. Drilling through the Ross Ice Shelf shows a very weak ice accretion rate of 0.02 m a^{-1} (Clough & Hansen, 1979).

The dimensions of the ice shelves are in part controlled by the geographical configuration of the embayments or coastal zone in which they develop (Swithinbank & Zumberge, 1965; Thomas, 1979). Often confining topography is required with the presence of submarine banks or small islands to act as 'pinning' points to hold-in the shelf. If such conditions are not met, an ice shelf would be quickly broken up by ocean currents, tide and wave action or even tsunamis. Break-up still takes place but only at the leading edge of the shelf where fracturing produces large tabular icebergs. These then float out into the Southern Ocean. It has recently been estimated that the annual iceberg production rate is 3–4 times higher than most previous estimates and could be in the order of 2.3×10^{15} kg (Orheim, 1984). The latest estimates of the annual mass of snow accumulating per annum in Antarctica are of the order of 2.01×10^{15} kg. This might suggest a weakly negative mass budget or one just in balance (given the rather gross assumption in the iceberg flux estimates).

Drilling into the ice sheet to considerable depth (2000 m) and airborne radar studies have both revealed a complex layered structure to the ice and in some places the presence of large 'lakes' at the bed (Robin, 1983). As snowfall scavenges macro- and microscopic particles, gases and other matter from the atmosphere and deposits them on the ice surface, progressive accumulation creates a time sequence of atmospheric and environmental conditions which can be studied from ice cores (Robin, 1983). Such investigations have yielded a rich scientific harvest regarding past climatic fluctuations (air mass patterns, temperatures, precipitation rates). Of current importance is the content of certain chemical species such as anthropogenic pollutions DDT, PCBs, Dieldrin, heavy metals.

Considerable effort is being expended to assess the likely stability of parts of the ice sheet, particularly that in West Antarctica, to warming and sea level effects induced by natural and anthropogenic (carbon dioxide) factors (Bentley, 1983).

Antarctic ocean and sea ice

Icebergs released from the continental ice mass come under the immediate influence of the Southern Ocean. Tracking of their drift by ships and satellites has measurably assisted in elucidating the pattern of currents around Antarctica (Fig. 2.8) (Techernia & Jeannin, 1983). Close to the continent the Eastwind Drift carries icebergs and sea ice westwards in a

Fig. 2.8. Principal elements of the oceanography of Southern Ocean. Shading depicts areas above 3000 m below sea level (therefore highlighting continental shelves and the mid-ocean ridge systems). The axis of the circum-Antarctic current (the Westwind Drift) and the Antarctic Divergence are shown. The short black arrows represent near shore currents, (after Zwally *et al.*, 1978).

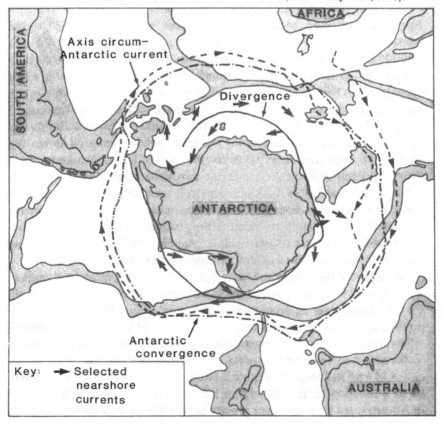

belt a few hundred kilometres wide and generated by the Coriolis deflection of katabatic winds draining off the ice sheet under the influence of high pressure. Flow is complex however with large and small-scale eddies which disrupt the current due to the effects of topography associated with the continental shelf. Further out, beyond the continental margin in latitudes of 40–60°S, the dominant current is the eastward-flowing Westwind Drift. This transports water at rates of 7–12 m s^{-1} around the continent and constitutes the dominant circumpolar current. It has been estimated that the total transport in the Westwind Drift is of the order of 130×10^6 tonnes s^{-1}. Between the two currents is a complicated shear zone where there is net transport to the south into the Eastwind Drift and net transport to the north in the Westwind Drift. It is termed the Antarctic Divergence and is a zone of upwelling water. Nutrient-rich deep water is brought up and provides a substantial basis for much of the food web of the Southern Ocean (Deacon, 1984).

The northward flow in the Westwind Drift (Ekman drift) brings about contact with southerly directed subtropical waters. Because of its temperature and density the colder, Antarctic surface water plunges beneath the warmer and less dense sub-tropical water in a zone at approximately 50 °S latitude termed the Antarctic Convergence. This feature is marked by a pronounced gradient in temperature and salinity. At depth, mixing takes place between these water masses and generates a separate and distinctive layer usually termed the Antarctic intermediate current. The freezing of the upper layers of the ocean immediately around Antarctica during the winter months creates dense cold masses of water that sink down the continental slope and generate a further water body – Antarctic deep water. At depth, the water spreads out across the ocean floor.

The very large size of the Southern Ocean and its distinctive physical composition and processes play critical roles in the Antarctic environment. It is possible, for example, to point to the resulting richness of the marine fauna, which sustained the vast whale population of previous centuries, and which today are the focus of developing economic interests.

One aspect of the Southern Ocean that also requires particular attention is the sea ice cover. At its maximum in the winter–early summer, Antarctic sea ice extends over an area of some 20 M km^2 (Fig. 2.9). Unconfined by surrounding land (in contrast to the geographical setting of the central Arctic Ocean) the sea ice is progressively broken during the austral summer by the action of waves and currents. It is reduced to something in the order of 3–4 M km^2 in February. Figure 2.10 illustrates the changing area of sea ice as a function of time through a typical yearly cycle. Therefore, most of Antarctic sea ice is first year growth only, with few regions of multiyear floes. These latter are found principally in the western Weddell Sea, and in the Bellingshausen and Amundsen Seas. The presence of ice in the ocean has the effect of inhibiting the exchanges of energy, mass and momentum between the sea and the atmosphere (Polar Group, 1980). These factors play a crucial role in modulating global climate (WWO/CAS-JSC-CCCO, 1982).

The melting of the sea ice and its surface snow cover in the summer releases considerable fresh water to the near-surface layers of the Southern Ocean. In winter, salinity is increased by fractionation as the surface freezes over and later as brine drains downwards out of the developing sea ice canopy.

Because of the inhospitable Antarctic environment it has proved difficult and unattractive to study the basic processes operating in the Antarctic pack especially during its growth in the winter. This has placed particular importance on satellite observations which yield all-weather, day–night

Fig. 2.9. Extent of sea ice with 15% concentration around Antarctica for the months of September (maximum) and February (minimum) averaged for the period of 1973–6 as obtained from ESMR data 5 Satellite, (after Zwally *et al.*, 1983).

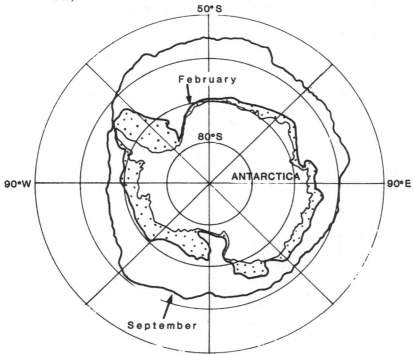

coverage. A major contribution by Zwally *et al.* (1983) has detailed the extent and distribution of Antarctic sea ice using the results of the ESMR (Electrically Scanning Microwave Radiometer) flown aboard the Nimbus 5 satellite. In the future, shipboard programmes are being developed in addition to new satellite missions, to investigate the interrelated glaciology–oceanography, meteorology and biology of the Antarctic Sea Ice Zone. In this regard, the Antarctic has fared poorly in comparison to the Arctic where military, strategic and economic factors have stimulated much research on sea ice during the last two decades.

Fig. 2.10. Variation in the annual sea ice cycle for the years 1973–6 from ESMR data. The figure shows areas covered by the 15% ice extent, regions of open water and the actual ice area (i.e. total ocean surface covered by sea ice excluding all open leads and polynyas; after Zwally *et al.*, 1983).

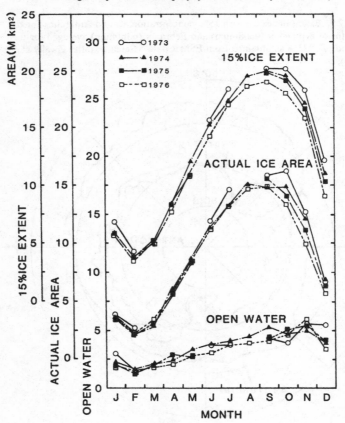

Fig. 2.11. Mean air temperature conditions in and around Antarctica for January and July (after Central Intelligence Agency, 1978).

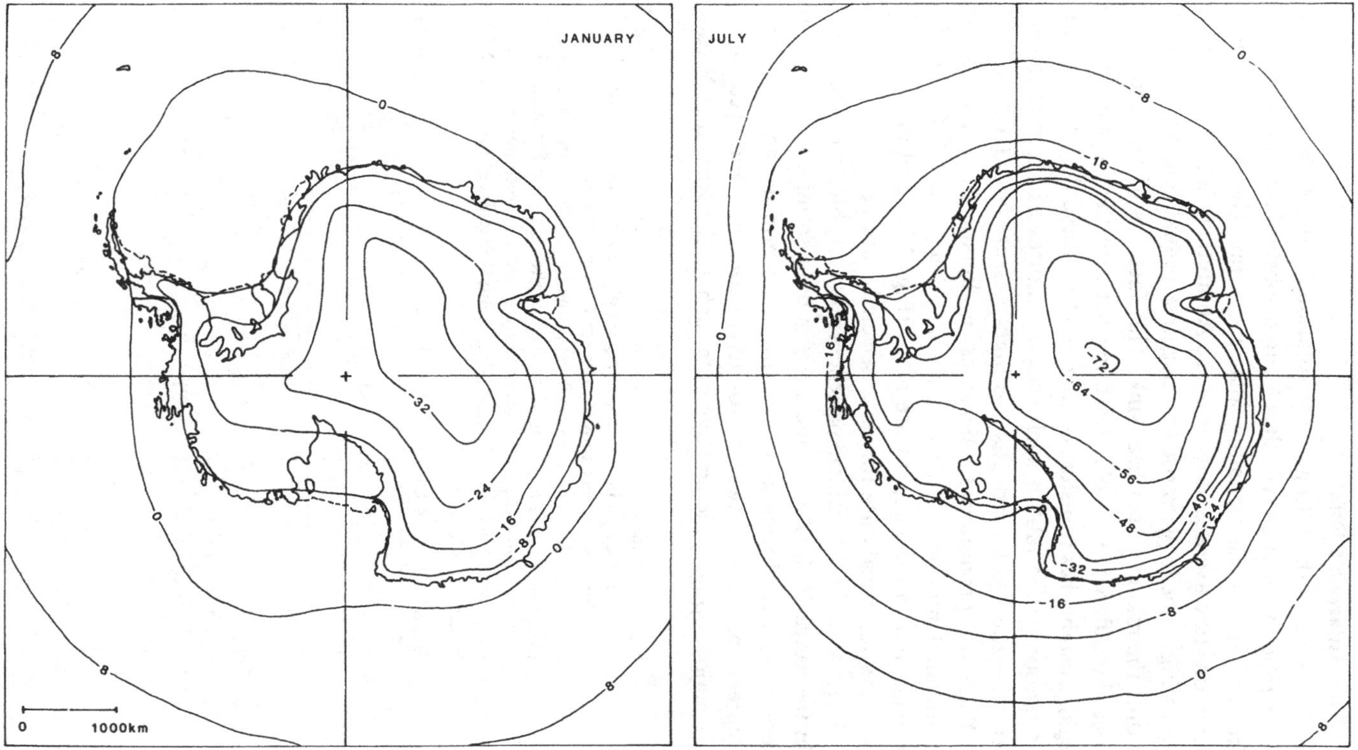

Antarctic Climate

Due to the tilting of the rotational axis of the earth by 23.3° to the ecliptic, Antarctica (which is located substantially above the Antarctic circle) receives little incoming solar radiation. The combination of high elevations (the average altitude of the Antarctic is 2000 m above sea level) with a high albedo gives rise to a negative annual heat budget. Antarctica is thus the major global heat-sink. This fact has widespread importance since the difference in solar warming between the tropics and the polar regions establishes transfers of energy which drive the global atmospheric circulation and establish such critical features as the intertropical convergence zone, polar troughs and high-pressure belts (Allison, 1983).

There are pronounced differences in climatic conditions between the continental interior, the coastal margin and the Southern Ocean. These are brought out in the temperature maps (Fig. 2.11) and in Fig. 2.12 which provides statistical data for three selected stations.

The Antarctic continent experiences relatively high barometric pressures (in the range 1030–40 mb) and there is only an infrequent occurrence of depressions. Cyclogenesis is, however, common in the Southern Ocean. Depressions move eastwards around the continent and occasionally enter the major topographic embayments such as the Ross and Weddel Seas.

Fig. 2.12. Climatic regime charts for three Antarctic stations (see Fig. 2.1). Vostok is located in the central region of the East Antarctic Ice Sheet at high elevation. Our planet's lowest surface temperature was recorded there in 1983 (−89°). Halley is close to sea level at the inner part of the Weddell Sea, whilst Deception Island is off the west coast of the Antarctic Peninsula. For the last two the mean daily temperature range is shown. Precipitation is given in minimum of water equivalent (after Central Intelligence Agency, 1978 with modification).

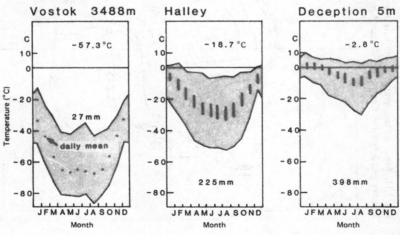

The very strong pressure gradients across the Southern Ocean generate a strong and unimpeded circulation system which gives rise to the extreme wind–wave character of circum-Antarctic waters.

On the Antarctic continent cold surface air layers lead to the widespread development of strong, stable inversions in the lower few hundred metres of the atmosphere. This phenomenon is probably one of the most distinctive elements of the Antarctic climate. The inversions are often 10°–15° 'deep' and favour development of katabatic air drainage – the flow of a cold, dense air layer close to the surface which descends the steepening slopes of the ice sheet (Parish, 1981). In coastal localities katabatic winds reach velocities in excess of 50 m s^{-1}.

References

I. Allison (ed.), *Antarctic Climate Research* (1983) Cambridge, UK, Scientific Committee on Antarctic Research.

J. C. Behrendt, D. J. Drewry, E. Jankowski, & M. S. Grim, 'Aeromagnetic and radio echo ice-sounding measurements show much greater area of Dufek Intrusion, Antarctica' (1980) *Science*, **209**, 1014–17.

C. R. Bentley, 'Crustal structure of Antarctica' (1973) *Tectonophysics*, **20**, 229–40.

C. R. Bentley, 'Crustal structure of Antarctica from geophysical evidence.' R. L. Oliver, P. R. James & J. B. Jago, (eds), (1983) *Antarctic Earth Science*, Canberra, Australian Academy of Science, pp. 491–7.

C. R. Bentley, 'The West Antarctic Ice Sheet: diagnosis and prognosis,' in Proceedings of Carbon Dioxide Research Conference (1983) *Carbon Dioxide: Science and Consensus*, Berkeley Springs, W.Va. IV-3-IV-32.

P. J. Cameron, 'The petroleum potential of Antarctica and its continental margin' (1981) *APEA Journal*, **21** (1) 99–111.

Central Intelligence Agency, *Polar Regions Atlas* (1978), Washington, DC.

J. W. Clough & B. L. Hansen, 'The Ross Ice Shelf project' (1979) *Science*, **203**(4379) 433–4.

C. Craddock, 'Antarctic Tectonics', R. J. Adie (ed) (1972) *Antarctic Geology and Geophysics*, Oslo Universitets forlaget, pp. 449–55.

M. M. Cresswell & P. Vella, (eds), *Gondwana Five* (1981) Rotterdam, A. A. Balkema.

I. W. D. Dalziel, 'The Evolution of the Scotia Arc' (1983) *Antarctic Earth Science*, Canberra, Australian Academy of Science, pp. 283–8.

I. W. D. Dalziel & D. H. Elliot, 'West Antarctica: Problem child of Gondwanaland' (1982) *Tectonics*, 1(1), 3–19.

F. G. Davey, K. Hinz, & H. Schroeder, 'Sedimentary basins of the Ross Sea, Antarctica' (1983). *Antarctic Earth Science*, Canberra, Australian Academy of Science, pp. 533–8.

G. Deacon, *The Antarctic Circumpolar Ocean* (1984). Cambridge University Press.

M. de Wit, *Minerals and mining in Antarctica: Science and Technology, Economics and Politics* (1985). Oxford Science Publications.

D. J. Drewry, 'Sedimentary basins of the east Antarctic craton from geophysical evidence.' (1976) *Tectonophysics*, **36** (1.3) 301–14.

D. J. Drewry, 'Surface of the Antarctic Ice Sheet', in D. J. Drewry (ed),

Antarctica: Glaciological and Geophysical Folio (1983a) Scott Polar Research
Institute, Cambridge, Sheet 2.

D. J. Drewry, 'Antarctic Ice Sheet thickness and volume' (1983b) *ibid.*, Sheet 4.

A. Ikami, K. Ito, K. Shibuya & K. Kaminuma, 'Crustal structure of the
Mizuho Plateau, Antarctica, revealed by explosive seismic experiments.'
(1983) *Antarctic Earth Science*, Canberra, Australian Academy of Science,
pp. 509–13.

P. R. James & R. J. Tingey, 'The Precambian geological evolution of the East
Antarctic Metamorphic Shield' (1983) *Antarctic Earth Science*, Canberra,
Australian Academy of Science, pp. 5–10.

E. J. Jankowski, D. J. Drewry & J. C. Behrendt, 'Magnetic studies of upper
crustal structure in West Antarctica and the boundary with East Antarctica',
ibid, pp. 197–203.

J. P. Kennett, 'Cenozoic evolution of Antarctic glaciation, the circum-Antarctic
Ocean, and their impact on global paleoceanography.' (1977) *Journal of
Geophysical Research*, **82** (27) 3483–860.

A. L. Kogan, 'Results of deep seismic soundings of the earth's crust in
Antarctica' (1972) *Antarctic Geology and Geophysics*, Oslo, Universitets
forlaget, 483–9.

R. G. Kurinin & G. E. Grikurov 'Crustal structure of part of East Antarctica
from geophysical data' (1982) *Antarctic Geoscience*, University of
Wisconsin Press, Madison, pp. 895–901.

N. F. McIntyre, 'The dynamics of ice sheet outlets' (1984) *Journal of
Glaciology*, **31** (108) 99–107.

O. Orheim, 'Iceberg discharge and the mass balance of Antarctica' (1984)
Iceberg Research, **8**, 3–7.

G. K. A. Oswald & G. de Q. Robin, 'Lakes beneath the Antarctic ice sheet'
(1973) *Nature*, **245** (5423) 251–4.

T. R. Parish, 'The katabatic winds of Cape Denison and Port Martin' (1981)
Polar Record, **20** (129) 525–33.

D. A. Peel, 'Antarctic ice: the frozen time capsule', *New Scientist*, **98** (1358)
476–9.

G. de Q. Robin (ed.), *The Climatic Record in Polar Ice Sheets* (1983),
Cambridge University Press, pp. 212.

P. D. Rowley, A. B. Ford, P. L. Williams & D. E. Pride, 'Metallogenic
provinces of Antarctica', *ibid*, pp. 414–19.

C. W. M. Swithinbank & J. Zumberge, 'The Ice Shelves' in Hatherton (1965)
Antarctica, pp. 199–220.

P. Techernia & P. F. Jeannin, '*Quelques aspects de la circulation oceanique
Antarctique revelés par l'observation de la dérive d'icebergs (1972–1983)*'
(1983). CNRS, Museum National de'Histoire Naturelle.

The Polar Group, 'Polar atmosphere–ice–ocean processes: a review of polar
problems in climate research' (1980) *Reviews of Geophysics and Space
Physics*, **18** (2) 525–43.

R. H. Thomas, 'Ice shelves, a review' (1979) *Journal of Glaciology*, **24** (90)
273–86.

M. R. A. Thomson, R. J. Pankhurst & P. D. Clarkson, 'The Antarctic
Peninsula – a late Mesozoic–Cenozoic Arc (Review)', in P. C. Oliver,
P. R. James & J. B. Jago (eds), *Antarctic Earth Science* (1983). Canberra,
Australian Academy of Science, pp. 289–94.

WMO/CAS-JSC-CCCO, *The Role of Sea Ice in Climatic Variations* (1982),
Geneva World Climate Programme, WCP 26.

K. J. Whitby, G. Rose & C. T. McElroy, 'Formational mapping of the Beacon

Supergroup type area with special reference to the Weller Coal Measures, South Victoria Land, Antarctica' in P. C. Oliver, P. R. James & J. B. Jago, (eds), *Antarctic Earth Science* (1983). Canberra, Australian Academy of Science, pp. 228–232.

H. A. Wright & P. L. Williams, 'Mineral resources of Antarctica' (1974) *U.S. Geological Survey*, Circular 705.

H. Yukutake & K. Ito, 'Velocities of P and S waves for drilling core rocks at Syowa Station, Antarctica' (1984) *Memoirs of National Institute of Polar Research*, Special Issue, 33, 17–27.

H. J. Zwally, J. C. Comiso, C. I. Parkinson, W. J. Campbell, F. B. Carsey & P. Gloersen, *Antarctic Sea Ice*, 1973–1976 (1983), Washington, DC, NASA SP-459, 206.

3

Scientific opportunities in the Antarctic

R. M. LAWS

Introduction

Although it is so remote, the Antarctic is a significant region amounting to nearly a tenth of the land surface of our planet and a tenth of the world's ocean. Formerly the keystone of the southern supercontinent Gondwana, it is now isolated from the other continents by a wide deep ocean, and all but 1 % of the land surface is covered by ice, over 4 km deep in places, a result of its polar position, which makes it the highest and coldest continent. The relative positions of the south geographic and geomagnetic poles have implications for geophysical studies. It has importance for meteorological and climatic studies, having a world-wide influence due to the fact that it is the strongest cooling centre of the global system. Being without an indigenous human population, Antarctica is still relatively unaffected by human activities. As such it can serve as a baseline for studies on global pollution of various kinds and the clear atmosphere confers advantages for certain research. A range of biological problems can be studied better there than anywhere else. A further advantage is the unique political situation and the absence of national boundaries under the Antarctic Treaty which allows large-scale international cooperation in science. However, because of the high costs of supporting research in such a remote and rigorous environment, there need to be very good reasons for conducting research there, rather than elsewhere. At the same time, the dependence of scientists on a common provision of logistic support tends to encourage interdisciplinary programmes.

A logical framework for my discussion would be to consider the research fields involved as we move upwards from the crust of the earth, roughly by way of the lithosphere – including geology, geophysics, plate tectonics; the cryosphere – glaciology, climatology and sub-ice shelf oceanography; the hydrosphere – oceanography; the atmosphere – meteorology and climatology; the ionosphere; and the magnetosphere – involving geospace studies. Finally, there is the biosphere, which crosses several of these boundaries – biology, physiology and ecology of living organisms. Although there are some cross-disciplinary connections (e.g. glaciology and climatology, oceanography and marine biology, solid earth geophysics and atmospheric and space physics), there is a wide diversity of specialised fields of science which provide opportunities for exciting and novel research of academic, strategic and applied value.

Geology and geophysics

In comparison with other continents, geological knowledge of the Antarctic is very limited. Serious research has been undertaken for little more than a quarter of a century and the limited rock outcrops, the hostile environment and logistic difficulties handicap workers. However, in a resource-hungry world, the subject is increasingly assuming importance, and the Antarctic Treaty is currently formulating proposals for a minerals regime.

Greater Antarctica is a Pre-Cambrian shield of igneous and metamorphic rocks bordered by the late Pre-Cambrian – early Mesozoic orogenic belt of the Transantarctic Mountains. The shield is structurally complex but important as an accessible and very ancient part of the Earth's crust, with ages in excess of 3500 Ma. The Transantarctic Mountains represent the major rock exposure in Antarctica and a natural geographical boundary between Greater and Lesser Antarctica. Lesser Antarctica is younger, of highly deformed and metamorphosed strata intruded by igneous rocks, consisting of a number of microcontinental blocks separated by subglacial basins, often below sea level.

The research fields involved include structural geology and plate tectonics, sedimentology, palaeontology and stratigraphy, marine geology, radiometric age dating, Pre-Cambrian and Phanerozoic history, radio-echo sounding, magnetometry, palaeomagnetism, seismic sounding and gravity surveys. Because only 1% of the rocks are exposed, outcrop geology has to be supplemented by extensive geophysical investigations of the vast intervening subglacial areas. These studies must be complemented by marine geophysical and marine geological research on the continental shelves and ocean basins, but the pack ice zone means that most of the shelf is poorly surveyed. This is partly compensated for by the fact that the rate of geophysical data acquisition at sea is much faster than on over-ice traverses. Airborne remote sensing is still more rapid, but is limited to radio-echo sounding of ice depths and airborne magnetometry. The use of synthetic aperture radar also has potential to help distinguish differing bedrock lithologies.

The continental drift theory of Wegener has now been accepted for many years but the basic mechanism was not demonstrated until plate tectonic theory was proposed in the 1960s. Nevertheless, the underlying causes are still not understood. There are several relevant problems of global significance that are being investigated. Of major interest is the break up of the original supercontinent of Gondwana. The present Antarctic plate is contiguous with seven other major lithospheric plates. A number of reconstructions have been made and the major uncertainties involve the Pacific margin from the Antarctic Peninsula to New Zealand. Studies may reveal important clues about the reasons for the break up and lead to improved understanding of major processes within the earth's mantle. Clues to the present biogeographical distributions involving the fauna and flora of the southern continents are being found as a result of palaeontological studies, particularly in Lesser Antarctica. The present Southern Ocean circulation has also arisen as a consequence of the break up.

Within present-day Antarctica, an important problem in global tectonics concerns the nature of the junction between Greater and Lesser Antarctica. What is the extent, structure and geological history of the microcontinental blocks of Lesser Antarctica, the largest of which is the Antarctic Peninsula? How are they interrelated and how do they relate to the craton of Greater Antarctica? What is the nature of the intervening sea and ice-filled basins with their potential as hydrocarbon reservoirs? Igneous rocks with metallic minerals of actual or potential value as resources are found in the surrounding continents (South America, Africa and Australasia) and in portions of Antarctica (e.g. the Dufek Massif and the Transantarctic Mountains). What is their tectonic and petrogenetic significance?

The overlap of the Antarctic Peninsula and southern South America is another problem in most Gondwana reconstructions and is related to the opening of the Weddell Sea. This has grown slowly over 160–170 Ma on the south of a spreading ridge separating the South American and Antarctic plates. The Scotia Sea has formed over the past 30–40 Ma as a complicated series of episodes of extension behind an island arc like the present South Sandwich Islands, and has overridden much of the northern flank of the spreading ridge separating the South American and Antarctic plates. The evolution of the Scotia Sea has been marked by successive collisions between the ancestral South Sandwich trench and that same spreading ridge. Studies using remote sensing (Gloria, Sea Mark or Sea Beam) would help to identify the original spreading centre and fore-arc fabrics, and possibly the collisions, to assess the dynamics. Coupled with multichannel seismic investigations, dredging, coring and heatflow studies, this should yield exciting results.

There are few examples in the world of the migration of a spreading centre into a subduction zone, but one such ridge crest–trench collision is being studied along the western margin of the Antarctic Peninsula, by geophysics and onshore geology. The subduction history of the Antarctic Peninsula also has wider potential for determining the rates of subduction-related processes, and there are resource implications arising from the possibilities of mineralisation processes and metallogenesis. Thus, metallic minerals in the Andes are related to subduction of the Pacific Ocean lithosphere and we need to study and test the possibility of a similar relationship to the Antarctic Peninsula.

This is one of the longest lived magnetic arcs which has been active for some 180 Ma. It is quite unusual among ancient arc systems in that fore-arc and back-arc terrains are all well exposed. This combination of good exposure and a long history make it one of the best regions on earth for studying the interactions of tectonic processes and sedimentation and the results will therefore be of global significance. Calc-alkaline magnetism related to subduction of the Pacific Ocean floor extends from at least the early Mesozoic to mid/late Tertiary. The record of simple geometry of subduction over the last 65 Ma is uniquely well defined. In the last period of Cenozoic activity extensional tectonics are associated with the opening of the Bransfield Strait as a marginal basin and alkaline volcanism in a northeast trending back-arc basin.

Studies of biostratigraphy and biogeography can make important contributions to understanding of the development of southern floras and faunas. Mesozoic floras and faunas provide evidence for an austral marine faunal realm. The rich molluscan fauna has stratigraphic significance and is relatively easy to study. The discovery of *Lystrosaurus* in the Transantarctic Mountains demonstrated the faunal links of the present southern continents in the Triassic.

Palaeobotanical investigations, including detailed studies of fossil tree rings, suggest that Alexander Island, for example, was set in a warm climate in the Mesozoic or that it has moved. Palaeomagnetic measurements have not been able to resolve this problem but more sensitive instruments (a cryogenic magnetometer operating near absolute zero) hold promise. Fossil marsupials discovered on Seymour Island are relevant to understanding the dispersal mechanisms for land vertebrates and indicate that a land bridge with South America continued into the Cenozoic; this agrees with the first formation of oceanic crust in the Drake Passage 30 Ma ago.

There is another major opportunity for the study of late Mesozoic and Cenozoic climatic history. Antarctica is a unique laboratory for such studies, including the onset of continental glaciation. Antarctic Bottom Water (ABW), for example, is a principal component of the Southern Ocean circulation; it is the main driving force of southern hemisphere thermohaline circulation and influences world climate and sediments even into the Northern Hemisphere. Its formation is intimately related to the development of the present Southern Ocean circulation and to Antarctic glacial history which can be studied by examining sediments laid down, at high latitudes, under its influence; this is best done by coring. (In some sectors, however, the Antarctic Bottom Water may in contrast be a force for erosion rather than the deposition of bottom sediments.) Stratigraphic control is so far restricted to diatoms, silicoflagellates and magnetic polarity. These investigations will give information on the sedimentary signature and on glacial/interglacial variability of Bottom Water production. The major source is the Weddell Sea (WSBW), but formation and downslope transport of WSBW has never been studied, although such work is now beginning; the international Ocean Drilling Programme (ODP) will be undertaking work in the Weddell Sea in 1986/7. There is uncertainty about the mode of formation by bottom melting of ice shelves; this is being studied elsewhere by glaciologists (see below).

As well as being preserved in the bottom sediments of the Weddell Sea, Upper Mesozoic and Cenozoic strata are exposed along the margins of this sea, including the Antarctic Peninsula. In addition to throwing light on climatic history, sedimentary processes, and the development of Antarctic biota, these strata, especially in the Weddell, Ross and Amundsen Seas, represent probably the single largest unexplored body of sediment in the world. Investigations are important for assessment of the hydrocarbons resource potential.

Glaciology and oceanography

Oxygen isotope data from ocean sediment cores suggest that the ice sheets or ice shelves or both developed 28–32 Ma ago. Deep ice cores to bedrock may be more than 100000 years old, and contain a unique historical record indicating how the ice sheets influence and react to global climate changes, and permitting the reconstruction of the Holocene climatic history of Antarctica. The history of the ocean and atmospheric climate and even clues to the planetary system (from planetary dust and meteorites) are also present. Impurities indicate past air temperatures, accumulation rates, volcanic and solar activity, sea ice extent, aridity, changes in atmospheric carbon dioxide, and ice sheet elevation. There are plans to drill to depths of 500 m at a number of sites. Because it is remote from industrial centres and population (mainly Northern Hemisphere) the ice sheet contains a valuable record of background global pollution levels, including organochlorines and heavy metals such as lead, zinc, copper and cadmium; these can now be measured at 10^{-12}g g^{-1} levels.

A recently developed field of research is due to natural processes that concentrate meteorites in certain parts of the ice sheet, where they are found in large numbers. This is brought about by meteorites being buried in ice accumulation zones and then transported by ice flow to reappear and be concentrated at the blue ice ablation zone. Together with cosmic dust and other impurities, these deposits are without parallel; with ages of up to 600000 years they are intermediate between ice cores and seafloor sediment cores.

Certain parts of the ice sheet may be less stable than others and liable to decay over time periods of as little as a century. Much of the ice sheet of Lesser Antarctica rests on bedrock below sea level and this, it has been pointed out, is probably an unstable situation. Small increases in sea or air temperatures could lead to a sudden decay of this ice sheet, producing a global sea level rise of 4–5 m, with serious consequences for low-lying cities and ports. For these reasons investigations of the interactions of ice shelves with ice sheets and the ocean are important. As inland ice flows towards the coast, it is channelled into ice streams and as the slope or driving force increases, the pressures lead to bottom melting so that it slides over the bedrock. Finally, beyond the grounding line, when the ice is floating, there is no friction, although there may be shear restraint at the sides. The transition zone between where the basal ice starts to melt and the grounding line is a crucial region for studying the processes controlling stability of the ice sheet. The position of the grounding line can be determined by means of sensitive tilt meters even over ice 2000 m thick and monitored over the years to establish any trend and make predictions. If there is not an abrupt transition between the land-based ice and floating ice shelf, there is a better chance of stability. Local grounding areas exert a restraining influence, even in the middle of an ice shelf.

The bottom melting of ice shelves is the least known term in the mass balance of the ice sheet. The interactions and heat exchanges between ice and underlying ocean are now being studied, using hot water drills to melt holes through an ice shelf (average 200 m) so as to place instruments to measure currents, temperature, salinity and oxygen isotope distribution as a function of depth. It is thought that there is a stable layer of freshwater beneath ice shelves which would drastically alter potential heat exchange with the ocean. It is also necessary to establish the oceanographic circulation under ice shelves, using moored current meters seaward of the ice front as well as the subshelf meters.

The shape and dynamics of the Antarctic ice sheet is controlled by precipitation (notoriously difficult to measure), upper surface temperature and melting at the base. At present it is not known whether the mass balance is positive or negative, because the basic physical behaviour is poorly understood. Satellite radar is altimetry and could measure the surface contours of the ice sheet. This, integrated with radio-echo measurements of the bedrock on which the ice is resting, makes possible accurate estimation of the volume of ice. Such estimates carried out, say every 10–20 years, would provide the best measure of the trend in the Antarctic ice sheet overall, smoothing out regional fluctuations. They would have an extremely important predictive value.

There has been a great deal of research on sea ice in the Arctic but very little as yet in the Antarctic. Antarctic conditions are different for formation and break up, being characterised by a centrifugal oceanic, short life pattern as opposed to the relatively land-locked confined conditions of the Arctic basin, which promotes accumulation of more multi-year ice. There is a very much greater seasonal flux shown by satellite observations and annual fluctuations also occur. The annual variation in sea ice extent represents a doubling of the size of the continent. These variations affect the albedo, and moisture and heat exchange between ocean and atmosphere, so sea ice is an important modulating influence in ocean/atmosphere interactions. In addition, large polynyas form within the pack ice zone, such as the Weddell and Ross Sea Polynyas. They vary from year to year and their effect on atmosphere and ocean circulation is little understood. A link between the extreme el Niño conditions off Peru in 1982 and the formation of a Weddell Polynya has been suggested.

Estimates of heat energy fluxes over the pack ice are few and unreliable and the role of ice associated diatoms in the seasonal decay of the pack ice (through absorption of solar radiation) should be investigated in summer experiments and taken into account in modelling studies. Other summer experiments should include ice melt rate, radiation balances and the structure of the pycnocline; winter investigations should cover freezing processes, convection and haline-driven circulations and gas exchange. In the open water polynyas study of heat, moisture and exchange of gas particularly in winter, would be valuable.

Other aspects of physical and chemical oceanography requiring further attention are the processes at oceanic fronts; the dynamics and thermodynamics of the Antarctic Circumpolar Current and its behaviour in the vicinity of island groups like South Georgia; the dynamics of the series of oceanic gyres associated with the ocean basins around Antarctica; and the coupling of the Southern Ocean with the world ocean. Large-scale modelling activities will be necessary to elucidate the systems and processes involved.

Meteorology and climatology

Time series of meteorological records provide raw data for climatologists who are concerned with assessing the inherent variability and any possible man-made perturbations of the earth's atmosphere. As discussed above, long-term records exist in the polar ice sheets and in the deep sea sediments which allow us to infer climatic changes over long periods. In this section we are concerned with weather and climate-related measurements in the neutral atmosphere, which as yet extend only over decades, as against periods of tens of thousands of years recorded in the ice sheets and millions of years in sediment cores.

The behaviour of the neutral atmosphere depends on the interplay of many complex energetic, dynamic and chemical processes. The ultimate source of energy for the circulation of the atmosphere is the flux of solar radiation, which is of greater intensity in lower latitudes. The earth's surface is heated directly, and there is also some lesser direct heating of the atmosphere. Transfer of energy between earth and atmosphere involves fluxes of sensible and latent heat as well as of infrared radiation. The Antarctic ice sheet, a high cold dome, and the surrounding pack ice, which effectively doubles the size of the continent in winter, are strong reflectors of incoming radiation. Thus Antarctica has a large radiation deficit, which has a major effect on the global circulation. The dissimilar distributions of land and water in the two hemispheres also lead to differences between them in atmospheric circulation and to different climates. One feature is that the mean air temperature in the Southern Hemisphere is about 2 °C lower than in the Northern Hemisphere.

After the March equinox, atmospheric cooling predominates in the Southern Hemisphere, and a winter polar stratospheric vortex forms due to the build up of thermal gradients between high and low latitudes. The westerly winds strengthen and a vast whirlpool of air, with a cold dense core over the pole, extends to about 50 °S at an altitude of 20 km and to the tropics at 40 km, with warmer air to the north. In contrast to the Northern Hemisphere, this vortex persists relatively undisturbed until it is disrupted by a final warming in October or November, well after the spring equinox. This is associated with a large-scale subsidence of the stratosphere and an increase in amounts of ozone. These in turn lead to the summer circulation – characterised by higher temperatures, anti-cyclonic easterly flow, decreasing ozone amounts and generally weak gradients.

In addition to their importance in understanding the general circulation and inter-hemispheric differences, studies of the Antarctic stratosphere have immediate practical applications. The ozone layer shields the lower atmosphere and life on earth from the effects of ultraviolet radiation, and there is concern about a possible decrease in ozone due to human activities, for example injections of oxides of nitrogen arising from operations of stratospheric aircraft and surface emissions of chlorofluorocarbons which result in increases of chlorine in the stratosphere. Long-term research in the clear Antarctic atmosphere, by optical methods, indicated no significant trends in ozone content from 1957 to 1973, but an unpublished study shows that, in recent years, the spring values of total ozone above the Antarctic have fallen drastically, and it has been suggested that this is directly related to the growth of inorganic chlorine in the atmosphere. The circulation has as yet shown little change. If confirmed, the implications may be serious.

Related measurements of infrared active trace species (e.g. water vapour, carbon dioxide and other minor constituents) in the troposphere and of ocean surface temperatures are being made from ships over latitudinal transects and by satellite overpasses. These studies are aimed at determining chemical 'sinks', particularly in the Southern Ocean (47% of freshly emitted carbon dioxide is taken up by the oceans) and investigating transport between the hemispheres. An important aspect of this work is to improve assessments of changes in climate due to the 'greenhouse effect' – a build up of gases introduced into the atmosphere by human activities that could cause an increase in temperature and a possible melting of the ice sheets.

An international Middle Atmosphere Programme (MAP), concerned with the altitude range 10–120 km, has an important polar component, as has the World Climate Research Programme (WCRP). The former is concerned with the energetics, dynamics and composition of this region and the different atmospheric layers. These studies include the nature of exchange processes between troposphere, stratosphere, mesosphere and thermosphere and possible effects of solar variability. The WCRP aims to improve understanding of climatic change, drawing upon evidence from meteorology, glaciology, oceanography and remote sensing, in the hope that some predictive capability can be developed.

Regular surface and balloon sonde (upper air) data are important for weather forecasting as well as for climatic research. Unfortunately, the network of Antarctic stations is very sparse, but developments of unmanned weather stations can be expected to improve the density of observations once the initial practical problems are overcome. Improved predictive ability is required, not just for Antarctic operations, but also for the other southern continents. Antarctic data are being used operationally for global weather forecasting. Satellites supplement the weather map based on ground and balloon observations by remotely sensing characteristics of the atmosphere, such as developing cloud systems.

In addition the balloon sonde results are important for interpreting other atmospheric experiments, and specific meteorological experiments are being planned to improve understanding of physical processes. For example, investigations of the stable nocturnal boundary layer which persists through the polar night might be linked with research on katabatic winds. An International Weddell Sea Project is planned, covering aspects of meteorological, oceanographical, and sea ice investigations. This will involve the remote deployment of buoys to measure standard meteorological parameters, reporting by satellite, and the direct involvement of remote sensing satellites.

Satellite imagery has opened up possibilities of studying the earth's surface from space. Landsat, in operation from 1972 onwards, gives information on rock outcrops, blue ice areas (associated with meteorite accumulation), sea ice and shelf ice distribution. Nimbus, also from 1972, employs microwave imaging to show sea ice distribution (it demonstrated the existence of a previously unknown major oceanographic anomaly, the Weddell Polynya, in the 1970s).

The ionised atmosphere and geospace

Antarctic research on solar terrestrial physics is focussed on the ionosphere and the magnetosphere. About 99% of all matter in the universe is in the form of plasma, an electrically charged gas in which each atom has lost one or more electrons and so has acquired a net positive electrical charge. Above some 70 km the upper atmosphere is also an electrically charged gas which formed under the influence of X-radiation and ultraviolet radiation from the sun. The plasma created is called the ionosphere, the density of ions increasing up to about 300 km and then decreasing in the topside ionosphere.

The magnetosphere is a volume of space in the vicinity of the earth, dominated by the earth's magnetic field, which determines the physical behaviour of the plasma. The earth's magnetic field is like that of a bar magnet lying at a slight angle to the earth's axis of rotation. The magnetic field lines connecting the Northern and Southern Hemispheres extend far into space, yet within the magnetosphere which shields the earth from the solar wind. This wind is a continuous supersonic stream of electrically charged particles, mainly protons and electrons, emanating from the sun; events on the sun produce gusts in the wind. The solar wind compresses the earth's magnetic field on the sunward or day side, but on the night side beyond about four earth's radii it is drawn out into a tail almost a million kilometres long, which always points away from the sun. The dimensions of the magnetosphere fluctuate as the pressure of the solar wind changes and influences its intrinsic physical processes. When some solar wind plasma penetrates the magnetosheath and mixes with plasma nearer the earth, a brightening of the auroral display occurs at 100 km. Magnetic substorms cause charged particles in the tail to move along magnetic field lines towards the earth. The ionosphere is bombarded by charged particles, particularly at the polar cusps and around the auroral oval.

The polar regions have advantages for such investigations, one of which is that this 'ionospheric television' can be seen best and continuously during the long winter polar night. Thus, the ionosphere is the lower boundary of the magnetosphere and is the interface, transition zone or coupling region between the neutral atmosphere and space. Beyond the topside ionosphere is the plasmasphere which co-rotates with the earth and extends up to about four earth's radii out into space, where, at the plasmapause, there is a sudden decline in the concentration of charged particles. This is also an important boundary between plasmas of differing origins, densities and temperatures. Inside, the plasma is relatively cool and dense; outside it is warm and tenuous. The ionosphere and magnetosphere, taken together, are now termed geospace.

The southern polar regions are better for studies of these regions and processes than the north because, as we have seen, the Antarctic is a continent on which have been placed permanent stations, whereas the Arctic is an ice-infested ocean. In addition the geophysical configurations differ. In the south the magnetic pole is displaced further from the geographic pole than in the north (respectively 23° and 11° of latitude). This larger asymmetry means that the angle between the magnetic polar axis and the solar wind undergoes greater diurnal and seasonal variations than in the north. Also, in the south more varied combinations of magnetic dip angle, field intensity and so on, and geographical latitude (which determines the input of solar ultraviolet energy) are possible than in the north. For example the stations Faraday and Dumont D'Urville are at a similar geographic latitude but the angles of magnetic dip at their locations differ by more than 30°. At the former the magnetic field lines lie in the plasmasphere; in the latter they run out into the tail of the magnetosphere.

Such extreme geophysical properties can be exploited by systematic observations and by special experiments. For example, an analogue of the physicist's key experiment is the detailed study of particular, significant events related to the level of magnetic disturbance, caused by the solar wind. The large displacement of the magnetic pole means that in the south the auroral oval and the projection of the plasmapause at the earth's surface traverse a very large range of geographic latitudes as the earth rotates and are therefore recorded by many Antarctic observatories. The extreme conditions over and near the Antarctic are particularly suitable for testing theories, especially of the F-region (the layer of maximum density in the ionosphere) wind theory and electric field theory. The F-region trough, a relatively stable region of reduced ionisation, is coupled with the plasmapause and it too is most evident during the winter night. The long winter night helps the definitive study of the relation of auroral and ionospheric phenomena.

Many deep-space phenomena can be studied from the ground in polar regions and a variety of techniques is used to study these phenomena. Scientists use ionosondes, swept-frequency radars, to measure the electron density profile of the bottom of the ionosphere and the height of the reflecting layers. More advanced sounders (AIS) that employ computer controlled radar to analyse ionospheric echoes digitally have recently been introduced and make possible much more sophisticated studies. Measurements by other radars would also open up new research possibilities. Magnetometers measure the fluctuations of the ionospheric current system; magnetic pulsations caused by hydromagnetic waves propagating along geomagnetic field lines can be used to estimate plasma density in the equatorial plane. Riometers (relative ionospheric opacity meters) measure radio waves from the cosmos. VLF (very low frequency) radio receivers are used to record whistlers mode signals and to study the electron density in the plasmasphere; whistlers are caused by lightning at the other end of a geomagnetic field line. Direction finding equipment can reveal details of the shape of the plasmapause and the positions of ducts in the plasmasphere. Magnetic conjugate point studies of such phenomena at each end of particular magnetic field lines are therefore very important. All sky cameras, photometers, and low light level T/V are used to study the aurora.

Balloon and rocket experiments are being undertaken to study the ionised upper atmosphere directly and orbiting satellites can measure the properties of the upper atmosphere and space plasma *in situ*. The International Solar-Terrestrial Physics (ISTP) Programme plans to deploy up to six spacecraft from 1990 onwards to examine the solar wind and various parts of geospace. Together with observations from rockets, balloons and ground observatories, including some in the Antarctic, plasma processes common throughout the universe can thus be studied in detail. The polar regions, particularly the Antarctic, are most important for the ground-based investigations.

Biological systems in the seas

The northern boundary of the Southern Ocean has not been acceptably defined but it lies north of the Antarctic Convergence. This is a physical boundary at an average latitude of about 50° S which marks the meeting of the northward-flowing cold Antarctic surface water and southward-flowing warmer water from the north. The true Antarctic biota are found south of this convergence. The uninterrupted circumpolar Southern Ocean is the windiest in the world and this in turn makes it the most turbulent. Primary producers in the phytoplankton and micro-plankton are therefore unable to maintain position in the euphotic zone to make optimal use of radiant energy from the sun. Thus, although nutrients are not limiting, levels of primary production are not significantly higher than in other oceans.

Despite this, the higher trophic levels – seabirds, seals and whales – are more abundant than in other oceans. This paradox is a singular feature of the Antarctic. Possibly it is to be explained by different pathways of nutrient cycling and the key position of an unusually abundant and long-lived pelagic crustacean, the Antarctic krill, *Euphausai superba*.

Ideas on the complexity and structure of Antarctic marine food webs and the rates of energy transfer are being revised. In studies elsewhere it is becoming apparent that there is significant primary production by nanoplankton and, together with bacteria and protozoans in the micro-plankton, this may represent an important alternative route of energy flow to that previously thought to be characteristic of the Southern Ocean ecosystem – extremely short food chains such as: diatom–krill–blue whale. The international programme Biological Investigations of Antarctic Marine Systems and Stocks (BIOMASS) has provided an unmatched basis for cooperative studies on these problems. Although it is due to end in 1987 it is to be hoped that the very successful international cooperation achieved under its aegis will continue.

A particular subsystem is the pack ice edge where the ecology of ice-associated algae, bacteria and protozoa deserves special attention. Winter studies within the pack ice zone are also important and great rewards will accrue to those research groups able to overcome the logistic problem of winter work in this environment.

Research on krill is important because of its central role in the ecosystem and because of its relevance for an expanding commercial fishery. Although this has not yet reached a significant level, the future of the higher trophic levels – squid, fish, seabirds, seals and whales – is dependent on the krill stocks being maintained at adequate levels of abundance.

Improved knowledge and understanding of the status of krill stocks and krill consumers is thus essential to the success of the Convention on the Conservation of Antarctic Marine Living Resources (CCAMLR).

But there are also many exciting research opportunities concerned with the behaviour, ecology, physiology and biochemistry of krill. These include large-scale studies of distribution, abundance and reproduction, the swarming behaviour and its relation to feeding. Experimental studies on swimming, feeding, oxygen uptake rates, swarming behaviour, moulting and activity patterns, metabolism and biochemistry, need to be undertaken.

All other levels of the Antarctic marine ecosystem provide exceptional opportunities for the advancement of knowledge of marine systems generally. A major lacuna is the cephalopoda, particularly the oceanic squids. We know from their occurrence in seabird, seal and sperm whale stomachs that they are abundant and ecologically important. The problem is to find ways of sampling and studying their populations. Having no swimbladder they cannot be detected by echosounders and, because of their speed and behaviour, they are difficult to catch in nets. A breakthrough in this area would provide very important opportunities. Fish are probably less important in the Southern Ocean than in the Arctic and have already been greatly overfished by commercial trawlers. Scientific opportunities perhaps lie more with the study of their physiology and biochemistry, and with their adaptations to the unusual Antarctic conditions. For example, most species show a reduction in red blood cells and these are lacking in the ice fishes such as *Chaenocephalus aceratus*. Another example is the experimental studies of the contractile properties of muscle fibres and ATPase activity which is of fundamental importance to the understanding of muscle physiology in general.

Antarctic seabirds have no fear of man and so they offer incomparable opportunities for the study of their behaviour, physiology and ecology. The rapid development of new methods of study (automatic nest weighing, activity recorders, time-depth recorders, radio or satellite tags and labelled isotopes for energetics studies) is leading to an explosion in knowledge. Coupled with the availability in several localities of large populations of known-age, banded individuals, this means that Antarctic seabird research is potentially at the forefront in these fields. In ecological studies we are especially concerned about the role of seabirds as predators on Antarctic marine resources, especially krill. At the population, cohort and general level a key question is: how is reproductive behaviour, reproductive success and longevity affected by age, experience, mate quality and limited resources (food and space)? The opportunity now exists to develop activity budgets, determine flight speeds and patterns, the duration of feeding trips, and the nature of feeding behaviour, by species. In energetic terms, fasting costs, flight and swimming costs can be estimated and related to food intake, both qualitative and quantitative, chick and adult growth, breeding frequency and survivorship. Similar opportunities are afforded by the Antarctic seals, including population structure and dynamics, density-dependence, social organisation and behaviour, the energy costs of foraging and onshore attendance by the lactating female, and the energetic efficiencies of milk transfer and pup growth. There is also an unusual opportunity to study factors associated with the recovery of the Antarctic fur seal from near-extinction to a position where this seal is now plentiful. It will also be possible to study the changes in the elephant seal population ecology following the end of commercial exploitation and the response of the world's most abundant seal, the crabeater, to the presumed increased availability of krill consequent on the reduction of the whale stocks. Similar opportunities to document the continuing increase in penguin populations exist, particularly in the Scotia Sea. However, with the recent decline in commercial whaling in the Antarctic, which is expected to end in 1988, the opportunities for whale research in the region are now very limited.

Studies of the benthos, part of the inshore marine ecosystem, have been in decline in recent years. Areas of particular interest still to be scientifically exploited are the sub-ice shelf communities, far from locations of primary production, the implications of the extreme environment – for example, slow growth and enhanced longevity of Antarctic benthos – for basic enzyme processes or protein metabolism. The long-lived species accumulate pollutants and may afford another opportunity for globally important baseline studies of pollution.

At the individual level, efforts should be made to exploit opportunities for experimental work on the adaptations of organisms to the extreme environment.

Terrestrial biology
The land area of the Antarctic is 14 M km², but no more than about 0.25 M km² is uncovered by ice and supports life. Inland waters cover a relatively small area, perhaps 5000 km², and many are permanently frozen.

The last major expansion of the ice cover came to an end about 10 000 years ago and glacial retreat since then has been followed by colonisation and the development of terrestrial and freshwater communities adapted to the extreme conditions. There are relatively few species of microbes, plants and animals compared with other regions and a high level of endemism, so ecosystem development is simple: energetic rates tend to be cold adapted and slower than in temperate systems. The Antarctic therefore provides ideal opportunities for investigating and modelling fundamental ecological processes – biogeochemical pathways, colonisation, succession, community development and cyclical changes.

The main habitats to be studied include the predominant fellfield, more advanced moss and vascular plant communities (including palynological profiles of peat banks in the sub-Antarctic islands and Antarctic Peninsula) and unusual habitats like the dry valleys and geothermally active areas; life is even found in snowfields in the form of algae and microorganisms.

The research promoting understanding of the ecosystems and their component organisms includes studies of the environmental variables. First, mechanical weathering of newly exposed rocks by freeze–thaw processes and by wind abrasion, chemical weathering and biological weathering by lichens. Secondly, microclimates are important, both in relation to weathering and soil formation (periglacial activity and particle sizes) and to colonisation. Temperatures, radiation, wind, soil temperatures, humidity and soil water tension can now all be measured easily with modern instrumentation. The microclimates are characterised by annual, seasonal and diurnal variations; light penetration of snow and ice and 'micro greenhouse' effects due to thaw and freeze may also be important.

Colonisation can be studied in relation to time since glacial recession by reference to known-age moraine terraces; within a small area a wide range of such conditions can be found along a transect for comparison. Experimental manipulation is possible. The critical factors appear to be temperature, moisture and nutrients. Such studies involve microbes, microflora, protozoa in soil, epilithic and endolithic colonisation. Opportunities for biochemical studies include investigation of the seasonal changes in soluble carbohydrate and amino acid content of coloniser microorganisms and the effect of freeze–thaw cycles on the release of carbohydrates from mosses and lichens. Wind, birds and sea transport all have a role in initial colonisation; the propagule bank can be assessed by culturing soil for potential colonists. There are opportunities for formulating principles involved in the present distribution and colonisation – on the offshore islands and the continent.

Survival depends on adaptations to the environment, and there are distinctive characteristics relevant to the Antarctic biota. Adaptations to desiccation and freezing involve cold hardiness and the synthesis of cryoprotectants, photosynthesis and respiratory physiology, including modification of carbon fixation rates, pigments, and enzymes for low levels of light, temperature and moisture. At higher levels, field activity, energetics and physiology (the feeding and growth of microarthropods and the physiology of water balance under these conditions) need to be studied. The principles involved can be related to other more complicated systems.

Antarctic communities have been very little, if at all, affected by man directly by perturbations or indirectly by pollution, which is minimal. These are relatively simple ecosystems in which the vegetation present is determined by climatic and geographical constraints and interspecific competition is slight because monospecific communities are common. (The biomass and productivity are often higher than in comparable Arctic communities). Such conditions are unusual in the world and facilitate the testing of hypotheses about biological strategies. It is possible to assess the likely effects of major or minor changes by manipulation of these simple ecosystems.

The absence of large herbivores has had a profound effect on the structure and functioning of these ecosystems. This results in a community with microherbivores, saprovores and a few micropredators, and the function of microorganisms is enhanced in such species-poor ecosystems. This unusual trophic structure has led to the development of ecosystems that are unique. They offer a possibility of studying the dynamics of food webs and constructing meaningful functional process models which may be profitably compared with the situation in, for example, species-poor hot deserts. The very low levels of man-made pollutants have important advantages for microbiological studies.

In addition to their academic interest, the likelihood of commercial developments at some time in the future calls for research into the effects of environmental impacts on these fragile ecosystems. The ability to predict the consequences of such disturbances is essential to the formulation of policy on commercial exploitation and ecosystem management. This calls for basic understanding of natural environmental variables and of growth, reproductive and survival strategies of at least the ecologically important species. Also needed is knowledge about the dynamics of colonisation, community development, ecosystem energetics and the functional interactions between the biota and the physical and chemical environment.

In addition there are unparalleled opportunities to investigate a wide range of biological problems that can contribute to the development of general concepts in ecology and environment physiology. Particularly important is the opportunity to test ideas on cold tolerance of poikilotherms and on survival mechanisms evolved in response to other stressful conditions. Studies of plant–water relations are likely to be fruitful because this is probably a critical factor in determining metabolic rates and productivity in the polar environment. The value of long-term studies of plant and animal communities is clear, and the extremely low density human population and the lack of any significant land use by man makes possible the dedication of land to monitoring sites.

The inland water bodies are a complementary part of the terrestrial system, influenced by events and processes in their catchment areas. These lakes have the added advantage that they are naturally closed systems, even more so when sealed by the formation of surface ice. They have even simpler ecosystems than the land and, like the terrestrial communities, are affected by fewer external influences than lakes in lower latitudes, so they may be expected to reveal basic ecological principles more readily. Each is unique and a wide spectrum of variation is imposed by progressive deglaciation, influencing the relative development of the ecosystems in their catchment areas, and natural eutrophication by birds and seals.

Human biology

Antarctic communities are isolated for long periods and many are small and provide a relatively limited age range of 'captive' subjects. This is a condition which can be exploited for a range of investigations into metabolism, nutrition, physiology, behaviour, psychology and virology (notably common cold research). The effects of isolation on behaviour can be studied over long periods. Nowhere else for example is it possible to examine the biology of national groups isolated from each other.

In addition they experience rigorous environmental conditions, both when working around some high latitude stations in winter and on scientific field work, when even summer conditions can be extreme at high latitude or at even moderate altitudes. The most extreme condition of prolonged severe cold plus anoxia due to altitude is most unusual elsewhere in the world. The effects of cold, of acclimatisation and the role of catecholamines are research topics that have been taken up.

Diving in polar waters in wet suits is another extreme condition often associated with a rapid drop in skin temperature, and a more gradual fall in core temperature; both may affect the diver's judgement and safety. Results are relevant to conditions associated with oil rig diving in northern waters, and perhaps at some time in the future in Antarctic waters.

Other extreme conditions, for example the day length regimes, offer opportunities to investigate behaviour, circadian rhythms, or endocrine physiology related to day length changes – for example seasonal changes in sleep cycles, or in melatonin production from the pineal gland. This accumulates during the dark period, is associated with mood swings, and is possibly dissipated under intense artificial light.

These situations provide opportunities for research not just of intrinsic value, but with applications to other conditions – hill walkers and mountaineers, fishermen, commercial divers, offshore oil platforms and other isolated populations – the effect of cold on performance, the relation of accidents to cold and the effect of light and darkness on performance.

Conclusion

This discussion has encompassed a very broad spectrum of research, ranging from problems below the seafloor out into space. There are however two unifying themes to Antarctic research: the simplicity and uniqueness of the problems. This starts with the political arrangements, described in other contributions, which make this 'a continent of science'. Because the Antarctic is internationalised without national boundaries or Exclusive Economic Zones (EEZs), international scientific collaboration is simplified and its extent is incomparable. This is made possible by the absence of any significant human population in the region and the lack of industrial activities, other than offshore fishing for krill, fish and whales.

Further, the absence of people and industry means that environmental impact and pollution is minimal – lower than anywhere else on earth – another unusual and simplifying factor. The geographical pattern of the Antarctic region is also uncomplicated and special – a high, domed, circular continent surrounded by a wide, deep ocean, well demarcated by a roughly circular physical boundary, the Antarctic Convergence, and characterised by relatively simple ocean current systems and weather systems.

This extends to the upper atmosphere and beyond, and the clear unpolluted air makes observation easier, as does the absence of other consequences of industrialisation like power surges. The configuration of the earth's magnetic field and the resulting magnetosphere also makes the Antarctic unmatched for geophysical studies.

The continent itself has a relatively uncomplicated structure; rock exposures are readily accessible and some of the best and simplest examples of island arcs and plate tectonic structures are found in the surrounding ocean. The ice sheet is by far the largest, and questions about its stability have global importance. It also provides excellent opportunities for the study, in ice cores, of past climates and there is a unique basis for the study of the global levels of anthropogenic pollutants like radioactivity, heavy metals and pesticide chemicals because of its remoteness from the industrialised regions of the world (unlike the next largest ice sheet in

Greenland). It is surrounded by fringing ice shelves, which are found nowhere else and afford additional opportunities to study ice dynamics, the flow law for ice, and Antarctic Bottom Water formation which can influence climate in the Northern Hemisphere.

Finally, the biological research, the very simple land and freshwater ecosystems and the predominance of non-specific communities, provide opportunities that do not exist elsewhere for understanding ecological processes. The ocean too has a relatively simple food web – dominated by the unusually abundant swarming crustacean, the Antarctic krill, as the key species – and there are exceptional opportunities to study it and its consumers. This has been the major theme of the international BIOMASS programme. The effects of the vast uncontrolled experiment associated with the overhunting and decline of major stocks of krill consumers has also provided exciting scientific opportunities. Other krill consumers, seabirds and seals, have a land or sea ice phase in their annual cycle which together with their lack of fear of man – another property which applies only to Antarctic wildlife – affords further unusual opportunities for research. Finally, there are opportunities to study small human communities in this stressful environment.

The factor to be balanced against all these advantages of simplicity and uniqueness is of course the remoteness of the Antarctic and the rigour and hazards of the environment. This makes Antarctic research expensive but increasing numbers of nations are setting up Antarctic research programmes. The content of these programmes indicates that the objective of the research is primarily basic or strategic (having a potential economic return in, say 25 years), rather than applied. So far the applied element in research has mainly been concerned with conservation and wise management of resources rather than directly with exploitation. It seems likely that this will continue to be the case for several decades at least.

References

I. Allison, *Antarctic Climate Research* (1983), Scientific Committee on Antarctic Research, Cambridge.

D. R. Drewry (ed.), *Antarctica: Glaciological and Geophysical Folio* (1983), Scott Polar Research Institute, Cambridge.

J. R. Dudeney, 'The ionosphere – a view from the Pole' (1981), *New Scientist*, **127**, 714–17.

J. Fuchs & R. M. Laws, 'Scientific research in Antarctica' (1977), *Phil. Trans. Roy. Soc. B.*, **279**, 1–288.

R. M. Laws (ed.), *Antarctic Ecology* (1984), 2 vols, Academic Press, London.

R. M. Laws, 'The ecology of the Southern Ocean' (1985), *American Scientist*, **73**, 26–40.

Oliver, James & Jago (eds), *Antarctic Earth Science* (1983), Australian Academy of Science, Canberra and Cambridge University Press.

G. de Q. Robin (ed.), *The Climate Record in Polar Ice Sheets* (1985), Cambridge University Press.

M. J. Rycroft, 'Atmospheric research in Antarctica.' In *Fundamental Studies and the Future of Science*, C. Wickramasinghe (ed.) (1984), University College, Cardiff Press.

Siegfried, Condy & Laws, *Antarctic Nutrient Cycles and Food Webs* (1985). Springer Verlag, Berlin.

Wolff & Peel, 'The record of global pollution in polar snow and ice' (1985), *Nature*, 313, 535–40.

Part II
The Antarctic Treaty regime: legal issues

4

Introduction

The four-way veto

Just as all roads lead to Rome so all agreements, recommendations and practices within the Antarctic Treaty system are referrable to and are explained by differing juridical positions on sovereignty. Four groups of state interests can be identified which adopt significantly different legal perspectives on the question of Antarctic sovereignty:

- Antarctic Treaty states which claim territorial sovereignty in Antarctica;
- Antarctic Treaty Parties which deny, or do not recognise, claims to territorial sovereignty and which make no claim of their own;
- Antarctic Treaty Parties which do not recognise any claim to Antarctic sovereignty but which reserve their own rights to make a claim in the future;
- states which are not party to the Antarctic Treaty regime but which deny claims to sovereignty on the ground that Antarctica is, or should become, part of the common heritage of mankind.

Each group of states has the political and legal power to protect its interests in a manner which has been described by Australia's Ambassador Brennan as a four-way veto. It is this power of veto which must be defused or avoided both within the treaty system, the negotiations for a minerals regime and the United Nations itself if the Antarctic Treaty regime is to survive.

The claimant states

Seven states claim territorial sovereignty in Antarctica: the United Kingdom, in 1908; New Zealand, in 1923; France, in 1924; Australia, in 1933; Norway, in 1939; Chile, in 1940 and Argentina, in 1942. The claims of Argentina, Chile and the United Kingdom overlap. Norway is the only claimant which does not define its claim with reference to coordinates through the South Pole – creating a pie chart – and its precise territorial limits remain unclear. The areas known as Ellsworth Land and Byrd Land are unclaimed. Argentina, Australia, Chile and New Zealand have direct interests in Antarctica which derive from their geographical proximity to the continent – weather, ocean currents, fisheries and defence strategies. Of the Northern Hemisphere claimants, Norway was concerned to protect whaling and sealing resources and the United Kingdom and France, their exploration and scientific activities. Claimants have historically asserted their claims to sovereignty on the basis of a variety of doctrines of territorial acquisition which are of more or less dubious legal validity in modern law. They include the Inter Caetera Bull of Pope Alexander VI dated 4 May 1493, the theory of *uti possidetis* (an intra-American rule of customary law of succession to states' rights) discovery, the sector principle and formal annexation. Today, claims to sovereignty rely upon the now-attenuated concept of effective occupation being the 'intention and will to act as sovereign, and some actual exercise or display of such authority'. The validity of these claims at modern international law may be doubted and, indeed, the question of validity itself begs the question whether the Antarctic continent is, in any event, amenable to territorial claims. It should, however, be recognised that serious territorial claims have been made, and acted upon, since 1908. It is probable that, had their validity been litigated, these claims would have been conceded according to principles of acquisition, at least until relatively recently. It is also arguable that the practices of states within the Antarctic Treaty system create some sort of title or preferred interest in Antarctic management.

Whatever the result of legal analysis, it suffices for present purposes to observe that international law, like all law, is dynamic, and that traditional concepts of territorial acquisition are now being overtaken by international political claims that the earth's 'common spaces' and resources are *res communis* or the common heritage of mankind. It is also legally significant that most states within the international community either do not recognise individual claims to sovereignty or have taken no informed position on the question. Only Australia, New Zealand, Norway, France and the United Kingdom reciprocally recognise their respective claims. Further it should be noted that not all claimant states are inflexible in their assertions of sovereignty. Norway has indicated that, in the interests of rational management in Antarctica, it might be willing to forego its claim. Others, such as Australia and the United Kingdom seem primarily concerned to promote the objectives of the Antarctic Treaty regime, thereby arguably prejudicing their assertions of full sovereignty. For the present, however, sovereignty is something of a trump card which has been employed to maintain a preferred position in any Antarctic negotiations.

The Treaty Parties which make no claim and deny assertions of sovereignty by others

These states number 23 and represent most political and economic systems and the bulk of the world's population. A significant exception are the states of Africa. South Africa's presence as a Consultative Party is seen by some States as a major flaw in the Treaty system. Non-claimant Parties' interests in Antarctica vary, as do their approaches to the sovereignty question. It is unclear what role they will play in the future. Some states might oppose the creation of any mineral resource regime based on sovereignty, others might reject the notion of a common heritage, and others might favour a 'wilderness' proposal and a total moratorium on resource development. Certainly it is hoped that these Parties, having a stake in the Antarctic Treaty regime, will understand the advantages of regional management within, rather than outside, the system.

Reservation of a right to claim

Only the United States and the Soviet Union deny existing sovereignty claims but reserve their rights, or bases, to make a claim in the future. It is ironic that these states far outstrip the claimants in present scientific and exploratory activities and would arguably have better legal title in Antarctica were they to claim it. The ideologically opposed positions taken by the Soviet Union and the United States in relation to the 1982 Convention on the Law of the Sea indicate their likely approaches to negotiation of a minerals regime within the Antarctic Treaty system.

The common heritage of mankind

The most articulate and active proponents of the view that Antarctica is, or should become, the common heritage of mankind are Malaysia, and Antigua and Barbuda. Importantly these states are not parties to the Antarctic Treaty or its related agreements. Malaysia, in particular, has employed the General Assembly as the international forum in which to argue that

> Antarctica as the common heritage of mankind requires a regime that is truly international in character...the exploitation of its resources must be carried out for the benefit of mankind.

For this reason, it is argued that negotiations by the Antarctic Treaty Parties of a minerals regime in Antarctica is exclusive and secretive. Those states favouring the concept of a common heritage in Antarctica do so in the context of the New International Economic Order and argue for the logical extension to Antarctica of the notion of a common heritage derived from outer space and the deep sea-bed. They believe that the Consultative Parties are a self-appointed and self-interested group of States which has no right to make decisions in Antarctica for the international community.

There is as yet no clarity or legal precision in the notion of a common heritage as it might apply in Antarctica; the phrase while employed by many states, disguises various interests and different substantive content. For the present, the ideal of a common heritage is a political aspiration which provides an impetus to reform within the Antarctic Treaty system. Indeed, there are indications that the early strength of these views has been weakened, partly because so many of the powerful and populous states are now within the Treaty system and partly because of moves to reform decision-making provisions, to improve the relationship with other international bodies and to make information more readily available.

The Antarctic Treaty system

Soon after negotiation of the Antarctic Treaty, Dag Hammerskjöld, Secretary-General of the United Nations, in an introduction to the 1960 United Nations Annual Report, called for a 'balance of prudence' to avoid new conflicts in the world. In many respects the Antarctic Treaty regime is the result of the exercise of restraint and a wider view of national and international interest than is frequently displayed. The Antarctic Treaty did not reflect an idealistic view of the sanctity of science or of a need to preserve the unique beauty of the Antarctic continent, so much as a necessary compromise to avoid rivalry between Argentina, Chile and the United Kingdom and the brooding dangers of Cold War between the United States and the Soviet Union.

The Treaty system does not constitute an international organisation with international personality in any accepted sense. It has no standing secretariat (although proposals have been made at the Thirteenth Consultative Meeting to establish one), and there is no central arrangement for the circulation of information or proposed measures. Rather, the system is decentralised and functionally oriented. It has been developed in an evolutionary way to meet particular needs as they arise. The Antarctic Treaty Parties simply bind themselves to meet

at suitable intervals and places, for the purpose of exchanging information, consulting together on matters of common interest pertaining to Antarctica, and the formulating, considering and recommending to their Governments measures in furtherance of the principles and objectives of the Treaty.

In practice, those Contracting Parties which are named in the preamble to the Treaty and certain acceding and appropriately qualified states, described as Consultative Parties, meet every 2 years at a conference hosted and organised by one of the Parties. Consultative meetings usually last 2 weeks, and the Thirteenth Meeting took place in Brussels in 1985. In the absence of a permanent secretariat, the host country assumes all administrative functions. The biannual conference is then supported by a range of subsidiary meetings including preparatory meetings, special consultative meetings and meetings of experts. This semi-permanent conference of 18 states has been the mechanism by which the Parties have negotiated Recommendations (which then become binding upon Consultative Parties) and other conventions linked to the Treaty itself. Over 150 Recommendations deal with tourism, environmental conservation, oil contamination and mineral exploration and, in 1964, the Agreed Measures for the Conservation of Antarctic Fauna and Flora were negotiated. The Antarctic Treaty Parties have separately negotiated a Convention for the Conservation of Antarctic seals in 1972 and the Convention on the Conservation of Antarctic Marine Living Resources (CAMLR) in 1980. Each of these Recommendations and conventions is linked to the Antarctic Treaty thereby forming a regime with jurisdiction over areas and issues which would not have been contemplated by those negotiating the Antarctic Treaty in 1959.

The regime remains, nonetheless, a weak and shadowy structure given to hortatory Recommendations rather than clear enforcement mechanisms. Recommendations of the Consultative Parties become effective only once they are adopted unanimously rather than upon the more usual basis of a two-thirds majority. They are not binding on non-Consultative Parties to the Treaty unless those Parties specifically accept them and, of course, Recommendations are not binding upon states which are not party to the Treaty.

The salient features of the Antarctic Treaty are well known. The continent is to be used for peaceful purposes only and all military measures or manoeuvres or the testing of any type of weapon is prohibited. All scientific investigation in Antarctica is free and the Parties agree to cooperate to achieve this objective through the exchange of scientific information, personnel and observations. Antarctica is the first nuclear-free zone in the world. All nuclear explosions are prohibited and radio-active waste material may not be disposed of within the Treaty area.

Critical to the success of the entire Antarctic Treaty regime is Article IV of the Antarctic Treaty which preserves the conflicting positions of claimant states, potential claimants and non-claimants. This much analysed and maligned article, has achieved what it set out to achieve – preservation of the status quo regarding sovereignty, thereby enabling the parties to proceed to regulate the Antarctic environment. Article IV has also provided a mechanism by which all subsequent conventions can be linked to the Antarctic Treaty itself. Perhaps the most notorious example of linkage is yet another Article IV, this time in CCAMLR which permits what has been described as a 'bi-focal' interpretation. These legal acrobatics enable each state to interpret its Treaty obligations in a manner most suited to its juridical position on sovereignty. Intriguing though these techniques are for the diplomat or lawyer, and justifiably criticised as they may be for a lack of objective legal content, these 'Articles IV' have facilitated the evolution of a framework for Antarctic regulation which is remarkable given the diametrically opposed positions of the negotiating Parties.

5

The Antarctic scene: legal and political facts

ROLPH TROLLE-ANDERSON

Antarctica of today is a continent of peace and cooperation. Eighteen countries with diverse political and economic systems successfully cooperate in managing all matters concerning the southern continent. A further 14 have expressed their political belief in the Antarctic Treaty system by adhering to the Antarctic Treaty. The last few years have seen an important increase in the number of new member states.

When one considers the potential for conflict in the Antarctic the situation of today is quite remarkable. In spite of widely differing conceptions as to territorial sovereignty, and undaunted by fluctuations in the global political climate over the last 25 years, the Antarctic Treaty Consultative Parties have unfailingly pursued the high goals set by themselves in the Antarctic Treaty. As will be seen from my following remarks, the achievements by the Consultative Parties in preserving the Antarctic from political and juridical strife are both considerable and, in their consequence, of benefit to all mankind.

In order to assess the political and legal facts in Antarctica today, it is necessary briefly to look back in time to the beginning of this century. I shall not endeavour to go into the details of the grand days of the Antarctic explorers of earlier times. It is of importance to keep in mind, however, that human enterprise has been significant on and around the southern continent for a very long time.

During the twentieth century, exploratory, scientific and commercial expeditions were carried out in the Antarctic by a number of countries. Seven of these countries – the United Kingdom, New Zealand, France, Norway, Australia, Chile and Argentina – made formal territorial claims to parts of Antarctica between 1908 and 1942. The claims of three of these countries (the United Kingdom, Chile and Argentina) partly overlap. One part of the continent, the area between 150 °W and 90 °W, remains unclaimed.

I shall not here try to analyse the legal grounds on which the territorial claims are based. Suffice it to say that the grounds are many, and that they vary, often combining principles such as discovery, occupation, geographic continuity and contiguity, formal acts of taking possession, legislation and performance of administrative acts. What is important to keep in mind, however, is the deeply rooted legal, political and emotional attachment to her territorial sovereignty in Antarctica which is felt by each of the claimant states. An understanding of the complexities of Antarctic politics of today is dependent on an appreciation of this attachment.

Also important, is an appreciation of the views of the so-called non-claimant states. Two of these states – USA and USSR – do not recognise the existing claims to territorial sovereignty and have avoided presenting their own claims, while at the same time maintaining that they have a basis for presenting such claims. The other non-claimants have neither recognised the claims or the bases of claims put forward by others, nor presented their own.

Such was the situation at the end of the Second World War. Non-recognition of claims by some, and inherent risks of conflict among the states with competing claims were the political realities of Antarctica. Against that background, attempts were made to initiate discussions with the aim of establishing some sort of international regime for Antarctica. The attempts failed, as did initiatives in 1956 and 1958 to put the question of Antarctica on the agenda of the United Nations.

At the same time as – and, to a certain extent, in spite of – this underlying disagreement as to the question of territorial sovereignty, international scientific cooperation, continued in Antarctica. Both separately and in cooperation, a number of countries sent out expeditions, carried out scientific investigations and established research stations. The apex was reached during the International Geophysical Year (IGY) 1957–8, where 12 countries – the seven claimants, the 2 superpowers, plus Belgium, Japan and South Africa – cooperated on the Antarctic section of this gigantic, worldwide project. These 12 countries established 60 stations for wintering on the Antarctic mainland ice and on islands in the Southern Ocean.

The successful international research cooperation during the IGY led to the establishment of a scientific cooperative organisation – SCAR, or the Scientific Committee on Antarctic Research, in 1958. The organisation consists of a number of specialist groups, with a representation from each of the member countries, as well as a number of scientific international organisations. In an annual national report to SCAR, the member countries give information about their research plans and activities, and the different specialist groups regularly arrange meetings at which information is freely exchanged.

The positive experiences from the IGY were important also in preparing the basis for negotiations on a more encompassing international treaty on the Antarctic. In 1958, the United States proposed to the governments of those states that had been active in Antarctica during the IGY, the negotiation of an international treaty to secure scientific cooperation in the area and to ensure that all of Antarctica would be used exclusively for peaceful purposes. The ensuing negotiations resulted in the elaboration of the Antarctic Treaty on 1 December 1959.

The Antarctic Treaty is a remarkable agreement. Through it, 12 countries with opposing views as to the question of territorial sovereignty agreed to cooperate freely in scientific investigation in Antarctica, irrespective of assertions of territorial sovereignty by some. In order to promote international cooperation in scientific investigation, the parties also agreed, to the greatest extent practicable and feasible, to exchange information regarding plans for scientific programmes, to exchange scientific personnel between expeditions and stations, and to exchange and make freely available scientific observations and results. At the same time, every encouragement was to be given to the establishment of cooperative working relations with those specialised agencies of the United Nations and other international organisations having a scientific or technical interest in Antarctica.

The Treaty furthermore set down that Antarctica shall be used for peaceful purposes only and effectively prohibits any measures of a military nature. Nuclear explosions and the disposal of radioactive waste material are prohibited. At the same time, the use of military equipment and personnel for scientific or any other peaceful purpose is not excluded, and each of the parties have the right to appoint their nationals as observers to carry out inspections everywhere in Antarctica, including inspections of all stations, installations and equipment, as well as of all ships and aircraft at points of embarking or discharging cargoes and personnel in Antarctica.

The cornerstone of the Antarctic Treaty is, however, its Article IV. Without that article, the Treaty would not have come into existence. Through it, the 12 parties could secure their main goals for the Antarctic – demilitarisation, denuclearisation and scientific cooperation – while at the same time not jeopardising their individual views on the question of territorial sovereignty. What Article IV does is to create the necessary legal framework for protection of each state's respective position on the question of territorial sovereignty.

Article IV provides that nothing contained in the Treaty be interpreted as:

- a renunciation by any party of previously asserted rights of or claims to territorial sovereignty in Antarctica;
- a renunciation or diminution by any party of any basis of claim to territorial sovereignty in Antarctica;
- prejudicing the position of any party as regards its recognition or non-recognition of any other state's right of or claim or basis of claim to territorial sovereignty in Antarctica.

This first section of Article IV thus deals with the possible consequences of the Treaty itself. The second part deals with the possible impacts of actions while the Treaty is in force. The second paragraph establishes that no acts or activities taking place while the Treaty is in force shall constitute a basis for asserting, supporting or denying a claim to territorial sovereignty in Antarctica or create any rights of sovereignty in Antarctica. The final sentence of the Article establishes that no new claim, or enlargement of an existing claim, to territorial sovereignty in Antarctica shall be asserted while the Treaty is in force.

It is on the basis of this ingenious formula of agreeing to disagree that the Antarctic Treaty Consultative Parties have managed to cooperate in Antarctic matters for more than 25 years. Through the effectiveness of this Article, Antarctica has remained the zone of peace which the signatories of the Treaty had hoped for. It is important to keep in mind, however, that Article IV has not resolved the questions of territorial sovereignty in Antarctica. It has removed the need for the parties to the Treaty repeatedly to reassert their respective views on the sovereignty issues and made it possible for these parties to cooperate peacefully in spite of their differences of view. The various claims to sovereignty in Antarctica still exist – as does the non-recognition of these claims. Article IV has worked so well, not by solving the issue of territorial sovereignty, nor by removing the possibility of conflict, but by the dual mechanism of removing the *need* to react to protect one's legal interests and of the appreciation by all parties that it would not be in their interest to push their different views to their logical conclusions.

The Antarctic Treaty did not intend to cover all aspects of human activity in Antarctica. Notably, matters connected with exploitation of resources were not dealt with. What the Treaty did do, however, was to create the necessary vehicle to deal with these and other matters of common interest. Through the cooperation that was established, and by way of the so-called Consultative Meetings, the Consultative Parties have succeeded in elaborating far-reaching measures for the protection, administration and management of Antarctica.

The Antarctic Treaty operates without the services of a secretariat. The smooth and proper functioning of the Treaty thus depends on the effectiveness of the Consultative Meetings. These are meetings where matters of common interest are consulted on, information is exchanged, and measures in furtherance of the principles and objectives of the Treaty are formulated, considered and recommended to governments. According to Article IX of the Treaty, such measures should include measures regarding the

- use of Antarctica for peaceful purposes only;
- facilitation of scientific research;
- facilitation of international scientific cooperation;
- facilitation of the rights of inspection;
- questions relating to the exercise of jurisdiction;
- preservation and conservation of living resources.

Antarctic Treaty Consultative Meetings – both regular and special – operate on the basis of consensus decisions. In practice this means that discussions about an issue continue until an agreement is found which all Consultative Parties can live with. It has been argued that this system of decision-making weakens the possibilities of reaching effective decisions – in view of the fact that decisions would always be based on the lowest common denominator. It would appear that the opposite might in fact be true. No decisions are forced through on the basis of a majority vote against the will of one or several Consultative Parties. The decisions that are eventually reached thus have the backing of all Parties thereby ensuring that these decisions will be implemented. On the basis of the diversity and scope of the recommendations and other measures that have in fact been agreed upon over the years, a case could also be made for the effectiveness and significance of consensus decisions – at least in the Antarctic context.

Thirteen Regular Consultative Meetings held at approximately 2-year intervals have taken place since the Treaty entered into force. In the course of these meetings over 140 recommendations have been adopted, covering all facets of human activity in Antarctica – with special emphasis on environmental protection. Other fields include *inter alia* meteorology, telecommunications, transport and logistics, facilitation of scientific research, identification of sites of special scientific interest, tourism, exchange of information, and a survey of the operation of the Antarctic Treaty system. Agreed Measures for the Conservation of Antarctic Fauna and Flora were adopted in 1964, and a Convention for the Conservation of

Antarctic Seals was signed in 1972. The most single important achievement is, undoubtedly, the Convention of 1980 on the Conservation of Antarctic Marine Living Resources – the so-called Krill Convention, or CCAMLR. The Convention builds on an ecosystem principle which views all marine living resources south of the so-called Antarctic Convergence (i.e. the approximate line where the cold waters from the Southern Ocean meet the waters from the surrounding oceans) as one system. This means that, in managing the resources one must look at the ecosystem as a whole, and that one must seek to maintain the ecological relationships among the different species – both those that are being harvested and those that are dependent upon the harvested species.

The CCAMLR is still in its early phases. It entered into force in 1982, and established its first protection measures at its third meeting in September 1984.

The Convention is significant in more than one respect. In addition to its ecosystem approach in management, it is also innovative in its establishment of a permanent Secretariat, with headquarters in Hobart, Tasmania. This is the first permanent body established under the Antarctic Treaty System. Legally and politically the most significant aspect of CCAMLR is that it was – once again – possible to agree on an important and far-reaching agreement in spite of the different views on territorial sovereignty. The parties succeeded in agreeing on an article in the Krill Convention which safeguarded their differing views in an acceptable manner.

The Consultative Parties are presently engaged in a series of negotiating sessions under a special consultative meeting on mineral resources. The main stumbling block is the lack of agreement on sovereignty. I shall not here attempt to enter into the intricacies of these negotiations. I should like to point out, however, that a satisfactory solution to the issue of territorial sovereignty is far more difficult to reach when one deals with minerals. In a minerals regime it will not be possible to circumvent the sovereignty problem through an 'Article IV' approach alone.

The Antarctic Treaty entered into force on 23 June 1961, having been ratified by all 12 original signatories. It is a common myth that the Treaty has a lifespan of 30 years, and that it will cease to exist in 1991. I should like to underline that this is not the case. What the Antarctic Treaty *does* provide for is the following. On the basis of unanimous agreement by all Consultative Parties, the Treaty can be amended or modified at any time. Such amendments or modification could in other words take place at any given period after the entry into force of the Treaty, without having to wait for 30 years. All that is needed is consensus by all Consultative Parties.

The misunderstanding concerning the 30 years validity period stems from a provision in the Treaty which would allow any Consultative Party to request a conference of all Contracting Parties to review the operation of the Treaty, 30 years after the entry into force of the Treaty. Since this took place in 1961 it would mean that any Consultative Party could request such a conference in 1991. Two main points should be kept in mind at this stage. One is that the Treaty could go on indefinitely, should no Consultative Party request such a Conference. The other is that, even if a Conference is called and amendments or modifications are approved by the Conference, these amendments or modifications will not enter into force without ratification by *all* Consultative Parties. Even after the 30 year period any amendment of the Antarctic Treaty thus demands acceptance by all Consultative Parties to come into effect. Without such acceptance an amendment would not enter into force and the Treaty would continue unchanged. The Treaty does, however, at this juncture open for any Contracting Party to *withdraw* from the Treaty. The precondition for such withdrawal is that an amendment or modification which has been approved by the Conference still has not entered into force 2 years after its communication to all Contracting Parties. In other words, any Contracting Party – be it a Consultative or Non-consultative Party – has the option to pull out of the Treaty should an approved amendment or modification not have been ratified by all Consultative Parties within a 2 year period.

Since the entry into force of the Antarctic Treaty in 1961, 20 more countries have acceded to the Treaty. Of these, six - Poland, the Federal Republic of Germany, Brazil, India, Uruguay and the Peoples' Republic of China – have obtained so-called Consultative status by demonstrating interest in Antarctica by conducting substantial scientific research activity there, such as the establishment of a scientific station or the despatch of a scientific expedition.

The Treaty thus today consists of 18 Consultative and 14 Non-consultative parties. The Treaty is open to any member of the United Nations. The last few years have seen a number of new accessions, such as the People's Republic of China, India, Hungary, Sweden, Finland and Cuba. All parties now have the opportunity to participate as observers at the regular Consultative Meetings and at the special Consultative Meetings on a minerals regime for Antarctica. This is a development which both shows the openness of Antarctic Treaty cooperation and the evolutionary character of the Treaty system. It is a development which has gone smoothly. The participation of the Non-consultative parties at the recent minerals negotiations in Rio de Janeiro underscores that point. Here, all

32 contracting parties to the Antarctic Treaty worked together, with meaningful participation for all in both plenary and in all working groups.

It has been charged that the Antarctic Treaty system is closed, and that the Antarctic Treaty Consultative Parties constitute some sort of club – secretly deciding on the fate of the last continent. Some countries have demanded that Antarctica – like the deep seabed and outer space – should be considered the common heritage of mankind and therefore administered by the United Nations on behalf of all mankind.

The parties to the Antarctic Treaty have opposed these views. A number of legal and political arguments have been presented. The Treaty parties have maintained that a comparison of Antarctica with the deep seabed and outer space is misleading. They have stressed that Antarctica is by no means a no-man's land. Active scientific, commercial and political activities have taken place in Antarctica during most of this century, and seven countries maintain claims to territorial sovereignty. Although such claims are not universally recognised, they nevertheless pose constitutional and political realities which cannot be overlooked. Finally, it has been pointed out that an encompassing international treaty as well as an extended cooperation system exist in Antarctica.

The parties to the Antarctic Treaty have, furthermore, referred to the openness of the Treaty, and urged states with an interest in Antarctic matters to accede to the Treaty. They have pointed to the ability to develop and to meet changing requirements that are the characteristics of the Antarctic Treaty system. They have stressed the singular achievements of the Treaty system in furthering peaceful cooperation in and peaceful uses of Antarctica. They have underlined that the Antarctic Treaty is a unique example of post-war international cooperation, engaging all parties in a constructive common undertaking – in spite of tensions and variations of the political climate among member states of other places in the world. Finally, they have warned against any attempts to break up or undermine this cooperation, pointing to the fact that it is built on an extremely fine and delicate political and legal balance which it would be impossible to recreate in the political climate of today.

6

The Antarctic Treaty system: a viable alternative for the regulation of resource-orientated activities

F. ORREGO VICUÑA

The Treaty system as an alternative to sovereignty and universality

The current debate about the various alternatives for the regulation of activities in Antarctica is not at all new. Ever since international law became interested in questions posed by polar exploration, with particular reference to the issue of resource exploitation – an interest that was already evident by the beginning of the century – the various alternatives conceivable in legal and political terms were brought into play.[1] Two basic proposals dominated the debate. The first purported to apply the traditional modalities of the organization of the modern state to Antarctica, extending the concept of sovereignty, with the necessary variations imposed by geography, climate and distance, to the polar regions.[2] The second dominant approach sought to negate such a possibility and to introduce, instead, forms of international organization that were generally of a global nature and on a world-wide scale.[3]

The approach relying on sovereignty alone proved not to be a viable alternative for the regulation of the activities of man in Antarctica. The reason for this was not strictly of a legal nature, and it was certainly entirely unrelated to the question of recognition or validity in international law. Rather it was one of political realities. Conflict and confrontation were constant features of the struggle between national sovereignties competing to become established, and such characteristics could not be the basis on which to found an effective Antarctic regime.[4] On the other hand, this approach did not properly take into account other relevant interests which were present in Antarctica, thereby also affecting the viability of this alternative. This is not to say that sovereignty does not have a role to play in Antarctica. On the contrary it does and an important one, but it cannot be the sole criteria for the organization of activities in this region.

Similarly, the opposite line of thought demonstrated serious short-comings when it was actively debated both at diplomatic and academic levels in the 1950s. The many suggestions that the continent be placed under the United Nations, trusteeship arrangements, condominium and various other forms of international organization, were shortlived.[5] The reason was that such an approach could not guarantee the absence of conflict and confrontation, on one hand, and the adequate satisfaction of the relevant interests including those related to national sovereignty, on the other.

It may be argued that the global approach never had the chance to be tried out in practice and that consequently it could only be judged in an intellectual perspective. That is quite true, but it is equally true that neither has sovereignty been intensively practised in Antarctica in any sub-stantive manner, most of its manifestations being of a symbolic nature. Therefore, in this case, the fundamental judgement is also of an intellectual nature. This is precisely the importance of the continuing international debate about Antarctica, including the present stage of the debate of the UN General Assembly,[6] in that it allows a better judgement about the viability of various ideas which from time to time have to be intellectually tested.

The complex set of arrangements that has come to be known as the Antarctic Treaty system is the natural outcome of the limits encountered by the two schools of thought in their effort to provide a model for the regulation of activities in the Antarctic. The Antarctic Treaty system is essentially a pragmatic formulation deprived of ideological connotations of any sort which enables it to sustain a continued process of compromise and adaption to the changing realities relevant to the Antarctic. Within this framework a number of important interests are taken into account and accommodated one with the other. These interests include national sovereignties, the position of non-claimant countries and the legitimate concerns of the international community. Above all, the system has been an effective guarantor of peace and stability in the region, thereby ensuring the fundamental basis of the development of successive regimes applicable to the activities in the region, which include environmental preservation, the management of living resources and the ensuing question of minerals.

Advantages of the Treaty system and general criticism

The views expressed by Australia in compliance with Resolution 38/77 of the UN General Assembly, have summarized well the advantages offered by the Antarctic Treaty system to the international community:[7]

(a) It is open to accession by any Member State of the United Nations, or any country which may be invited to accede with the consent of the Consultative Parties – it is thus as universal as the interest of States in Antarctica;

(b) It is of unlimited duration and establishes Antarctica as a region of unparalleled international co-operation in the interests of all mankind;

(c) It is based on the Charter of the United Nations, promotes its purposes and principles and confirms Antarctica as a zone of peace; it is, in fact, the only effective, functioning nuclear-weapons-free zone in the world today;

(d) It excludes Antarctica from the arms race by prohibiting any measures of a military nature, such as the establishment of military bases and installations, the carrying out of military manoeuvres or the testing of any types of weapons, including nuclear weapons, and forbids the dumping of nuclear waste;

(e) It encourages and facilitates scientific co-operation and the exchange of scientific information, which is made available for the benefit of all states;

(f) It protects the natural environment of Antarctica, including the Antarctic ecosystem;

(g) It provides for a comprehensive system of on-site inspection by observers to promote the objectives and to ensure compliance with the provisions of the Treaty;

(h) It has averted international strife and conflict over Antarctica, *inter alia*, by putting aside the question of claims to sovereignty in Antarctica, thereby removing the potential for dispute.

To this enumeration one could well add a number of other points which refer specifically to the various regimes set up under the Treaty system.[8] It is particularly noteworthy that none of these regimes confers rights on the member countries, but only impose obligations upon them. The regulation of activities in Antarctica is normally achieved through an agreed abstention by member countries from adopting measures or otherwise engaging themselves in steps which are considered inadvisable from the point of view of the environment or of the management of the resources involved. It is also important to note that such obligations do not affect the freedom of other members of the international community, except to the extent that the pertinent rules might become part of customary international law or that the treaty system could qualify as establishing an 'objective' territorial and managerial arrangement.

However impressive the achievements of the Treaty system might have been – a point stressed time and time again by the Consultative Parties – the fact remains that such a system has come under heavy fire from critics at diplomatic and academic levels of debate. In a recent contribution by Ambassador Zain-Azraai the criticism is well summarized in the following terms: he explains that one fundamental problem relates to the assertion of the Consultative Parties:

> that they – and they alone – have the right to make decisions pertaining to Antarctica ('exclusive'), that these would cover all activities in Antarctica ('comprehensive') and that these are not subject to review or even discussion by any other body ('unaccountable').

The Australian views referred to above have described the suggestions made during the 38th session of the General Assembly to revise or replace the Antarctic Treaty in the following manner:

> The main arguments for change appear to be that the Treaty system is anachronistic and discriminatory, that claims to sovereignty should be put to one side as a form of colonialism, that the system is secretive, that it is exclusive and controlled by developed countries, that it should be replaced by a universal regime, and that the benefits derived from the exploitation of Antarctic resources should be shared as the 'common heritage of mankind'.[10]

I should have preferred to explain the Antarctic system on its own merits, relying on the positive aspects that it encompasses and not doing so with a reference to a debate that has many negative connotations. Nevertheless since these critical arguments have been put forward, the merits of the system can also be explained by replying to and assessing such views.

A system of experienced and responsible participation

The question of participation in the Antarctic Treaty system, with particular reference to the decision-making procedures, has been one of the central points of criticism directed against these arrangements.[11] It has been argued, more specifically, that such arrangements involve a two-tier structure within which only the Consultative Parties have the opportunity to decide upon the administration and policies of the system. The resulting discussion, it is further argued, would be detrimental to the role of non-consultative parties and to the interests of the international community. The alternative of universal participation or a broadly expanded one is proposed as a solution.

Neither the Antarctic Treaty nor any of the regimes associated with it have established a closed system of participation. On the contrary, it is a well known fact that participation is open to all the members of the United Nations. The number of acceding states speaks for itself. Furthermore, participation in the decision-making procedures can and has been periodically enlarged, either by means of the new countries attaining the status of Consultative parties or of the increasingly flexible provisions on institutional participation that the regimes on resources have incorporated.

It is one thing, however, to qualify for participation in decision-making after having demonstrated a substantial interest in Antarctic activities, and an entirely different one to claim such participation as a 'right'. While the first approach is welcomed by the Treaty system, the second can be seriously questioned. When the first draft of the Treaty was discussed at the 1959 Conference on Antarctica, it provided for an automatic right of participation in the Consultative Meetings for every acceding country. It was Chile who proposed that a Consultative status different from that of a single accession was needed.[12] This was not done to discriminate against anyone or to affect legitimate interests of other countries, but only to ensure that the countries undertaking such a responsibility would be those really active in the continent.

This criterion was justified then and it still is today. If the arrangements consist mainly in the duties of abstention from the given activities or modalities, it is logical that only those participating in such activities should be entitled to participate in the management of the respective regime.[13] Moreover, it is reasonable that countries acceding to such activities will also be allowed to join in the management as is presently the case. This is not at all new as far as international law is concerned.

The system is obviously not 'exclusive' since it is open-ended to the extent of qualifying new states. It could be more fairly described as 'selective' in that the criteria for participation in decision-making are based on experience and expertise. While the idea of expertise has also been the subject of much criticism,[14] it is neither illegitimate nor incorrect. On the other hand, there is a practical necessity to ensure that the activities undertaken in Antarctica will be properly managed, since otherwise the consequences could be disastrous. There have already been serious difficulties with the installation of new stations in Antarctica because of the lack of the necessary experience.

The present characteristics of the Antarctic Treaty system do not mean that participation cannot be improved. Various measures were adopted on the occasion of the Twelfth Consultative Meeting with regard to the participation of observers from governments and international organizations. The subject will be debated further during the next Consultative Meetings. Similarly, the various alternatives for participation and accession are presently being discussed during the negotiation on mineral resources. Measures can also be implemented to facilitate the participation of countries that have not had access to the necessary experience, including joint research programmes, provision for infrastructure, joint ventures and the like. Some countries have already greatly benefited from this approach. The improved participation in the system will mainly benefit the parties of the Antarctic Treaty that have not yet attained the status of Consultative Parties, since there is some merit in the argument that these countries, although having the same obligations as the Consultative Parties, do not have the same functions.

Participation in the system has been and will continue to be enlarged, but this will be a gradual process linked to the criteria of activity in the continent and related areas. No such thing as a drastic change in the system is likely to take place, because the system has proved to be effective in relation to the values that it was entrusted to safeguard and develop.

Secrecy and publicity in the Treaty system

The accusation that the system is 'secretive' is usually linked to the question of participation.[15] While there is some merit in this view, two clarifications are appropriate. The first is that the system has not relied so much on 'secrecy' as on 'privacy', which is a rather different proposition. Governments and scholars who have wished to have access to the discussions or documents of the Consultative Parties have been able to do so on an informal basis or even by means of collections available in public libraries. While it is true that this information has not been available on a widespread basis, it is also true that neither has it generally been kept secret.

The second clarification is that this privacy has not been designed to keep the world community in a state of ignorance, but it is due to an entirely different reason. It would be most difficult for Consultative Parties to make public the nature of the compromises that need to be worked out in order to reach agreement on matters that are highly sensitive for the position of claimants and non-claimants alike, or on other matters that would result in adverse reactions by public opinion or their own governments. The effect of publicity would be to paralyse the capacity of the system to evolve and adapt itself to new realities. This phenomenon is not different from that which led the procedures of the Law of the Sea Conference to avoid records and publicity, and rely more on private and informal meetings. In any case, the dissemination of information is also one of the aspects that is currently being improved and broadened.[16]

The interrelated nature of Antarctic activity

Why the Antarctic Treaty system has evolved from having a limited scientific and peace-keeping purpose, enshrined in the 1959 Treaty, to being a comprehensive arrangement, which includes living and mineral resources, is not difficult to explain. All the activities taking place in Antarctica are closely bound together because of their very nature, and all of them have an effect on the values protected by the Treaty. Could, for example, peace and cooperation be guaranteed under the Treaty, or the environment be duly protected, if the resource-orientated activities were left unregulated or regulated in an inefficient manner? It seems that such a situation would be self-defeating for the objectives of the Treaty. This is why the system has the important ability to anticipate the new needs arising because of potential additional activities in the area, thereby preventing any danger of conflict or mismanagement. The Agreed Measures, the Seals Convention and the Living Resources Convention have been set up before the respective activities have taken place or acquired an intensive pace. This is also why the regime on mineral resources is being negotiated with a sense of urgency, thereby ensuring that if any significant discovery is made the rules will already be in place. Otherwise, the danger of conflict would certainly be extreme.

One other reason should not be forgotten in this regard: from the point of view of claimant countries, the territorial sovereign has the right to develop and regulate all the activities taking place in its territory, however comprehensive these might be. It can also choose to do so by means of agreements with other countries active in the area, which is the approach followed by claimants in Antarctica. As these are valid claims in international law, and as the respective titles have been perfected, the role of sovereignty in the Treaty system should not be overlooked, although, as has been said, it is not the only or single controlling factor.[17]

The Treaty system and the rights of states under international law

The existence of the 'right' of the Consultative Parties to regulate activities in Antarctica has been questioned in the current debate about the Antarctic Treaty system.[18] This line of reasoning was based on the wrong premise, because it assumes that such a right is equivalent to a privilege and that those benefiting from it constitute a self-appointed group. In other words, the idea of usurpation by Consultative Parties is the underlying thought of this point of view, although the expression has not actually been used.

In addition to the fact that the Treaty system does not strictly confer 'rights', but rather imposes obligations, it is important to understand that any right which the Consultative Parties may exercise is no different from the normal rights which all states have under international law to organize their activities in given areas or matters. The only limits that states have in this regard are that their agreements should not be contrary to the principles and purposes of the Charter of the United Nations. The Antarctic Treaty and related instruments are certainly not contrary to the Charter, and have moreover been specifically agreed to so as to further such principles and purposes. Why then should the countries most active in Antarctica not have the right to organize their activities in a lawful manner? Such a conclusion would certainly be a case of adverse discrimination in international law.

The parallel with the United Nations Conference on the Law of the Sea has frequently been made in support of the argument that universal participation in those negotiations was not related to the expertise of countries in given matters like seabed mining, but was based on interest and natural justice and this should also be the case in Antarctic affairs.[19] The comparison can also be stated in a different way: because all countries make use of the oceans, participation was naturally universal in the Law of the Sea negotiations, including those on seabed mining, which were conceived in a comprehensive manner; likewise, participation in Antarctic negotiations, including those on mineral resources, is related to those countries making use of the continent or related areas, which was conceived in an equally comprehensive manner. Whether such participation is universal or limited depends purely on the number of countries undertaking activities in the area, as this would also have been the case with the Law of the Sea had not all countries made use of the oceans.

If what is termed the 'right' of Consultative Parties is not recognized, implicitly there is an affirmation that such a 'right' appertains to someone else, and the international community is generally suggested for this purpose. The claims to universal participation would not then be purely a matter of interest but of right, because interest has to be positively expressed by means of activities or other practical manifestations which are not apparent in this case. If this is so, the questioning could be reversed and it might be asked: who gave the international community that right? And on whose authority? The burden of the proof then becomes more difficult since the international community does not have either the record of activity or the element of sovereignty to support such a right or title, and natural justice is not sufficiently specific in international law.

This is why the claim of the Antarctic Treaty system's accountability before the United Nations or any other body is not well founded, since it would suppose that such a body is superior or has a better right or title. Furthermore, the argument of accountability entails a power to revise international treaties that cannot be explicitly found in the Charter of the United Nations.[20] It will be remembered that precisely because of the problems that arose under Article 19 of the Covenant of the League of Nations – which my country experienced directly in the 1920s – the revision of treaties does not appear in the Charter and we could certainly not accept this form of indirect revival of an obsolete provision. The implications of the thesis of 'accountability' for the principle *Pacta Sunt Servanda* were clearly stated at the last United Nations debate, in that 'states that are not parties to a binding treaty ought not to be able, through the United Nations, to call into question the obligations of States parties to the treaty'.[21]

Many forms of cooperation can be developed between the Antarctic Treaty System, the United Nations and other international organizations, a process that is well under way. In fact, the Antarctic system can be reasonably viewed today as an international organization in itself without having reached a great degree of institutionalization, but which does appear to be distinctive in the evolution currently taking place. Is it necessary for the purpose of this cooperation, to choose an approach that involves the concept of subordination, which is inherent in the idea of accountability, thereby making it unacceptable? Might it not be preferable to think in terms of the relationship which ASEAN, the OAU and many other organizations have with their sister institutions throughout the world, including the UN family?

Universality and quasi-universality as alternative approaches: limits of the argument

Those countries that have questioned the so-called right of the Consultative Parties to regulate activities in Antarctica had to offer an alternative model. The idea of a universal model for Antarctica came once again to the fore as shown by the 1983 and the 1984 discussions of the UN General Assembly. The experience of the Law of the Sea Conference was specifically referred to in a continuous manner, and in particular the concept of the common heritage of mankind was repeatedly suggested as the appropriate model for application in this particular context.

Most of the Antarctic Treaty Consultative Parties have supported the concept of the common heritage of mankind in the seabed regime and the Moon Treaty. The case of Antarctica, however, is entirely different, because of two reasons that have been expressed time and time again: the first is that the Antarctic, unlike the seabed, has been subject to claims of sovereignty, and the second is that it is also the subject of specific legal arrangements under international law, unlike the other areas mentioned.[22] The concept of the common heritage cannot be extended beyond the specific regimes which have accepted it, unless an agreement can be found to the contrary. To date one has not been forthcoming.

During the Law of the Sea negotiations, Chile proposed that the concept of the common heritage of mankind be declared a rule of *Jus Cogens*, which by definition involves an element of customary international law.[23] This was done, however, with specific reference to the seabed and Article 136 of the Convention and not to any other source. It would have been inconceivable for my country to contradict the Antarctic claim by means of any other interpretation of that proposal.

Besides the universal approach, a second line of thought has also been apparent in academic literature and the United Nations debate. This other approach proposes to recognise given aspects of the Antarctic Treaty system that are considered positive, such as peaceful use, demilitarization and prohibition of nuclear tests. This approach also demands changes in the resource regimes, particularly the regime on mineral resources and changes in participation, decision-making and administration of the system.[24] The basic criticism of the system is maintained but the idea of a radical and total change is no longer put forward; in its place a more selective or restricted target for modifications is being given greater emphasis. This other approach is frequently presented as a procedural question, as the main task is to find a neutral forum for discussion of the matter without prejudgment of any position.

While it is always positive for some aspects of the Treaty system to be properly valued by this type of reasoning, the situation is more complicated. Firstly, there is no such thing as a purely procedural question at the United Nations, and the mere fact of establishing a Committee for the discussion of Antarctica would obviously involve a questioning of the Treaty system in order to deal with the issue recognizing a parallel mechanism under the United Nations. Secondly, and more importantly, the fact that some states have been explicit in their view that the Treaty system is a transitory or provisional arrangement while the process of total internationalization takes place,[25] means that the end result is definitely prejudged. Furthermore, the diplomatic record clearly shows that some of the milder criticisms of today were quite stormy only a year ago. In addition to the above, and irrespective of intentions or procedures, the criticisms that have been raised in regard to the Treaty system are not shared by the Consultative Parties. This renders the idea of major changes difficult to achieve.

The central point is that renewed pressure on the Treaty system has been correctly perceived by the Consultative Parties as an effort to upset the balance of the two basic historic approaches, reintroducing the idea of universalism as an alternative model. Whether this idea is explicit or implicit, whether it is more or less universal, or whether it is instant or delayed the end result affects the essence of the present system of limited international cooperation. This is precisely what is not acceptable to the Consultative Parties and other Treaty parties.

Given the effectiveness of the system and the combined power of the countries participating in it, no alternative is viable without the consent of the Antarctic countries. Just as the idea of total internationalization did not prove feasible in the pre-1959 suggestions, neither does it today. It is not realistic to try to repeat the Law of the Sea experience in a totally different legal and political situation and in a substantially changed international environment. Many changes can be justified in the Antarctic Treaty system and many are positively needed. Changes should, however, be the result of gradual reform, increased participation and a strengthened process of Antarctic cooperation.

The current debate about Antarctica has not been at all futile. It has once again proved the limits of internationalization and, more importantly, has, as a result, greatly strengthened the system itself. It can confidently be said that the importance attached to the system by the Treaty Parties is greater today than at any other time[26] not because of the existence of an external threat but because the case put against the Treaty system has not proved convincing. Hence the positive contribution of this system to the regulation of activities in Antarctica has emerged with greater clarity.

Endnotes

1 See generally the forthcoming book on the issue of Antarctic mineral resources by the author of this paper, Chapter I.
2 Gustav Smedal, *Acquisition of Sovereignty Over Polar Areas* (1931). Oslo: Norges Svalbard of Isharvs Undersokelser.
3 See generally Philip C. Jessup & Howard J. Taubenfeld: *Controls for Outer Space and the Antarctic Analogy* (1959). New York: Columbia University Press.
4 Francisco Orrego Vicuña, 'Antarctic conflict and international co-operation' (1985) In *US Polar Research Board: Workshop on the Antarctic Treaty System*, pp. 28–43.

5 For a discussion of various proposals for the internationalization of Antarctica, see C. Wilfred Jenks, *The Common Law of Mankind* (1958) Chapter 8: 'An international regime for Antarctica?'

6 UN General Assembly, 38th Session, Official Records, First Committee, 42nd–46th meeting (28–30 November 1983) A/CI/38/PV 42–46. Also 38th session, Official Records, First Committee, 50th–55th meeting (28–30 November 1984), A/CI/39/PV 50–55. See also Report of the Secretary-General on the Question of Antarctica, A/39/583 (Part I, 31 October 1984) (Part II, Vol. 1–3, on *Views of States*, 29 October 1984 and other dates).

7 Australia, *View of States*, UN Doc., A/39/583 (Part II) 29 October 1984, at 85.

8 See generally Francisco Orrego Vicuña, *Antarctic Resources Policy* (1983).

9 Zain Azraai, this volume.

10 Australia, *Views of States*, *op. cit.*, 7 at 86.

11 See n. *supra*.

12 Oscar Pinochet de la Barra, 'La contribucion de Chile al Tratado Antartico' (1985) 89–100. In Francisco Orrego Vicuña *et al.* (eds), *Politica Antarctica de Chile*, at 97–98.

13 United Kingdom: Statement on the question of Antarctica, UN Doc. A/CI/39/PV. 52, 30 November 1984, at 26.

14 Zain Azraai, *op. cit.*, n. 9, 182–3, 187.

15 A summary of the arguments made during the current UN debate can be found in Peter J. Beck, 'The United Nations and Antarctica' (1984) 22, 137, *Polar Record*, 137–144.

16 Recommendation XII-6, on the 'operation of the Antarctic Treaty system'.

17 On the question of title to Antarctic territory in international law, see Gillian Triggs, 'Australian sovereignty in Antarctica' (1981–2) 13, *Melbourne University Law Review*, Part I at 123–58, Part II at 302–33.

18 Zain Azraai, *op. cit.*, n. 9 at 183.

19 *Ibid.*, at 187.

20 Ian Brownlie, *Principles of Public International Law*, at 622. Oxford: Clarendon Press.

21 United Kingdom, Statement *op. cit.*, n. 13 at 33–5.

22 Rainer Lagoni, 'Antarctica's mineral resources in international law' (1979) 39, 1–37 ZACRV, at 34–5.

23 Doc. FC/14, 20 August 1979.

24 Evan Luard, 'Who owns the Antarctic?' (1984) 1174–93 *Foreign Affairs*. Also Zain Azraai, *op. cit.*, n. 9, 185–8.

25 See, for example, Christopher Pinto, 'Comment' (1984), in Rudiger Wolfrum (ed.): *Antarctic Challenge*, at 164–8.

26 Christopher Beeby, 'The Antarctic Treaty system as a resource mechanism: non-living resources', *US Polar Record Board*, *op. cit.*, n. 4, at 164.

7

The relevance of Antarctica to the lawyer

HAZEL FOX

Antarctica is unique in its isolated location (990 km from the southern tip of South America), size (one-tenth of the surface of the globe), permanent ice-cap (covering 98% of the continent) with consequent extremes of climate, and absence of permanent human habitation. Do such unique characteristics, stressed by the explorers and scientists who know the region, make law unnecessary? Regulation is required where a people grows in number beyond family and tribal constraint and exchange and communication with other groups of people take place. With the few thousand scientists at present in Antarctica and their logistic support, provided in part by personnel of the armed forces – both disciplined groups under their home state laws – there is at the present time, apart from some conservation and communication measures, little need for the apparatus of legislation, courts and law enforcement as it exists in the modern state.

What relevance, then, has Antarctica for the lawyer? Probably little, at the present time, for the practitioner in one particular national system. Material for the comparative lawyer and private international lawyer is equally scanty.

But, if the absence of an indigenous population dispenses with the need for laws to preserve internal order, the size of Antarctica and its untapped resources in an international community increasingly aware of its finite limits, has produced a conflict of interests between states. The reconciliation and regulation of such conflict of interests falls squarely in the field of the public international lawyer. Antarctica, therefore, and the legal issues which it has presented, ever since its explorers received state backing, is of direct relevance to the international lawyer. The handling of the issues can be considered in the four stages in which they have presented themselves.

In the first stage (that of the formulation of the claims to title of the seven states who by discovery, occupation, contiguity, inherited right or geological affinity assert territorial sovereignty over areas in the region) the skills of the international lawyer were required to arrange the peculiar types of state activity which the unique character of Antarctica produced into the classical categories governing title to territory. The United Kingdom's application to the International Court of Justice in 1955[1] with regard to sovereignty in its Antarctic possessions well illustrates this stage. It was fairly easy work for the competent lawyer, though it certainly presented some novel aspects, as in the development of the sector concept.[2]

The second stage was a much greater intellectual challenge. The group of claimant states was joined by non-claimant states, some making no claim but interested in participation in the area and two, the USA and the USSR, who did not recognise the existing territorial claims, had not presented their own yet asserted they had a basis for presenting such claims. A formula had to be provided to satisfy the diverse interests of these claimant, non-claimant and non-recognising states so as to enable them to enjoy freedom of science and information in the region, to preserve it for peaceful uses and to ensure its demilitarisation and denuclearisation. Article IV – the freezing of claims – was the lawyer's device which achieved these goals.

The third stage, which is still in process, is the fleshing out into legal regulation of the principles of the Antarctic Treaty of 1959 to cover the growing variety of activities which technological developments are making possible in Antarctica. General measures regulating marine living resources and tourism have been agreed and regulation of mineral resources is under discussion. In this third stage it has also become necessary to fit the special regime of Antarctica into the general scheme of international law relating to the making of international norms, allocation of jurisdictions and use and management of resources.

This coordination of special law with general rules of international law is likely to produce more conflicts of state interests as involvement in the Antarctic increases. The difficulties of resolving it satisfactorily are likely to take the international lawyer into the fourth stage of legal thinking – an abandonment of old legal methods based narrowly on state sovereignty and territorial jurisdiction in favour of new concepts to accommodate the conflicting demands. These four stages of legal thinking in connection with Antarctica provide ample material for the international lawyer. Legal developments in Antarctica illuminate the structure of international law and the international community as a whole. Their relevance is direct and cannot be dismissed as esoteric arrangements, made for some remote icy region.

Many problems still require legal elucidation in respect of each of the four stages. The rest of this paper will identify some of them and indicate the legal issues which arise.

The first stage

Although legal articulation of the territorial claims was worked out in some detail in an exchange of notes between the governments concerned prior to 1959 (Argentina, Australia, Chile, France, New Zealand, Norway, United Kingdom), no attempt has been made to assess the strength of the claims at the present day judged by classical law standards. Nor has any attempt been made on the basis of state activity in the area to reconcile the overlapping claims of Argentina, Chile and the United Kingdom or to determine the status of the unclaimed sector as *res nullius, res communis* or as in some way subject to inchoate title of the states who already have historic claims to the region. For third states not party to the Antarctic Treaty or on the assumption that the Antarctic Treaty provisions may at some time not govern the situation, such a study of recent state activity in the zones occupied by states in the area of overlap and in the unclaimed sector has direct relevance. If pressure from United Nations' involvement increases, such a study might usefully remind third states of the strength of some of the Consultative states' positions in classical law and also reveal, on the same basis, the degree to which the 18 Consultative states have earned priority of recognition by effort expended in the area.

The second stage

Thinking in terms of the first stage may well be thought over-academic in light of the Antarctic Treaty regime and the present commitment to it by 32 states representing all parts of the world, all the permanent members of the Security Council, India, Japan and certain substantial Third World countries. Concentration on the second stage may be more profitable and there is still much legal thinking to be done in respect of this stage. The scope of the general principles of the Treaty have by no means been thoroughly worked out. This is particularly true of the geographical scope of the treaty and the principles contained in Articles I, II, III and IV relating to demilitarisation, exchange of scientific information and the freezing of territorial claims.

The geographical scope

The geographical scope of the Treaty raises definitional problems: of where territory ends and sea begins (made more difficult in Antarctica by the presence of land and sea ice);[3] of where the Treaty area ends and areas governed by general international law begin. Article VI of the Treaty offers some guidance by providing that the Treaty shall 'apply to the area south of 60° S latitude including all ice shelves' but without prejudice to the rights of any state under international law with regard to the high seas within that area. In respect of marine living resources the parties to the Treaty have temporarily overcome these definitional difficulties. For purposes of the conservation of such resources, they have extended the Treaty area to the Antarctic Convergence (which enables areas of sea considerably north of 60° S latitude to be included) and adopted an approach of 'bifocalism'. This solution of 'contrived ambiguity' and the relevance of modern developments in the law of the sea such as the continental shelf and Exclusive Economic Zone (EEZ) are fully examined by Dr Triggs in Part III of this volume.

Article I

The general provision of Article I(1) which provides that Antarctica shall be used for peaceful purposes only and prohibits any measures of a military nature is qualified by Article I(2) which permits the use of military personnel or equipment for scientific research or any other peaceful purpose. At the present time a large number of Antarctic bases are supplied by the armed services and some bases have been set up and maintained by military personnel. Does Article I represent a qualified variant on the normal state of complete demilitarisation, waiving the objection to military character in respect of personnel or equipment, or is it a general, and the only practicable, model for any demilitarised regime?[4] What provides the military character of any activity? Is it the presence of weapons in immediate readiness for aggressive use, the establishment of bases in strategic locations with telecommunications, the training of men and testing of equipment under extreme conditions or the use in scientific research of men who are members of the armed forces of a state? Since the distinction between combatant and non-combatant is easily blurred, is the additional and major limitation on activity to be found in the condition in Article I(2) that military personnel and equipment must be engaged in a peaceful purpose, whether of scientific research or some other pursuit? The adoption of the Treaty regime ended the hostile naval patrols which occurred prior to 1959. The open employment and cooperation of military personnel of different states on scientific and other peaceful purposes, coupled with inspection, ground or aerial, being available to all areas of the Treaty including stations, provides an example of peaceful cooperation of government forces to be emulated elsewhere. In time of war or international tension, however, the peaceful presence of military personnel in strategic bases may quickly be converted to military purposes, as events in the recent Falklands War illustrated. Although the Falklands War did not spread generally to Antarctica, it began when a naval supply boat on a routine Antarctic trip brought Argentine nationals to raise the flag in South Georgia. Argentinian military personnel in 1977 took possession of Thule Island in the South Sandwich Islands (though British military personnel forcibly expelled them in 1982), only 96 km (60 miles) north of the Treaty area and well within the Antarctic Convergence as defined by the Convention on the Conservation of Antarctic Living Resources.

Articles II and III

Article III provides for free exchange of scientific information. The term is not defined but is linked to 'the freedom of scientific investigation in Antarctica and co-operation as applied during the International Geophysical Year [IGY] in Article II. It clearly does not cover investigation of all kinds. The limitation to scientific activities actually carried on during IGY 1957/8 is not today observed in that many other types of scientific investigation are pursued. The draft Convention on Mineral Resources currently distinguishes between the commercial development of mineral resources and the preliminary stages of exploration and prospecting, including the use of remote sensing, drilling or dredging for the purpose of samples. Elucidation is required whether information obtained in these preliminary stages falls within Article III and is freely available or whether it is protected by the ban relating to confidentiality of data which appears in the current draft of the Minerals Convention,[5] whereby operators are not required to exchange or make freely available data obtained from Antarctic mineral resource activities. As prospecting is an activity which is likely to be commercially practicable in the immediate future, this uncertainty requires resolution.

Article IV

Article IV, whereby territorial claims to Antarctica are frozen, of course, presents the most intriguing conundrum to the lawyer. Considerable problems of legal interpretation arise with Article IV and throughout it has to be borne in mind that it represented and still represents a political compromise of delicate balance. Article IV(1) provides that nothing contained in the Treaty shall be interpreted as a renunciation of previously asserted rights, claims or basis of claim to territorial sovereignty in Antarctica or so as to prejudice the position of any Contracting state as regards its recognition or non-recognition of such rights, claims or basis of claim. From the wording it is not clear whether Article IV(1) provides an evidentiary bar to reliance by any Contracting party on the Treaty provisions as changing the prior legal position or whether it goes further positively to impose obligations on a Contracting state to refrain from acting during the operation of the Treaty in accordance with its previously asserted position. Obligations are more clearly created for the period of the duration of the Treaty by the wording of the second paragraph of Article IV. Article IV(2) provides that no act while the Treaty is in force shall constitute a basis for a claim to territorial sovereignty in Antarctica and also bars the asserting of any new claim or enlargement of existing claim while the Treaty is in force.

The Treaty is not a bilateral agreement, as in the case of the Argentine/ UK Exchange of Notes incorporating the Joint Statement of 1971 in relation to the Falklands Islands.[6] Nor one, as in that Exchange of Notes which operates between states, all of whom claim identical rights of territorial sovereignty. Any obligation assumed is, therefore, not neces- sarily reciprocal. By Article IV a state may make a different concession according to its position as claimant, non-claimant or non-recognising state and the consequent obligations differ. Thus, it may be argued a non-recognising state merely undertakes during the operation of the Treaty not to make protests [and possibly, if Article IV(1) connotes obligation, not to repeat previous protests]. In exchange a claimant Contracting state will undertake not to assert present (or past) territorial sovereignty nor to treat such non-repetition of protests as prejudicing the non-recognising state's legal position. The non-uniform nature of the obligations is increased in respect of those states which accede to the Treaty at a date later than its entry into force or for states who undertake similar obligations by becoming a party to other Antarctic Conventions such as that on Marine Living Resources.

The non-reciprocal nature of obligations complicates the legal effect of Article IV. One line of analysis is to examine the limits of the Article's freezing effect by party, area, subject-matter and time. This involves consideration whether rights and claims to territorial sovereignty over areas in Antarctica can continue to be made under the classical law system against states who are not parties to the Treaty, in international forums outside the Treaty area, in respect of matters not covered by the Treaty or at a time after 1991 when, on failure to amend the Treaty a Contracting

Party may be free to withdraw (Article XII). Another line, which comple-
ments the first, is to examine the juridical nature of the freezing provided
by Article IV. Does it, as with a 'without prejudice' clause in civil litigation
or the interim agreement made between Iceland and the United Kingdom
as discussed by the ICJ in the *Fisheries Jurisdiction* Case,[7] create a second
regime of rights and obligations which exists side by side with the original
suspended regime? For third states and international institutions does the
original regime continue to have full legal significance? Is the purpose of
Article IV a temporary moratorium to permit settlement of the conflicting
claims it covers? Can it continue indefinitely if no settlement is achieved
or feasible?

Another line of analysis is to study the implications of Article IV and
the consequent conduct of the contracting parties during the operation of
the Treaty as constituting continued maintenance of the sovereign rights
and claims, despite the express language of the freezing clause. Thus, does
not the very presence of Article IV and its repetition in the Conventions
on Seals and Marine Living Resources and proposed inclusion in the
arrangement for minerals, amount to a recognition that such rights and
claims exist and require account to be taken of them? Does not the
restriction of the activities of seven of the Contracting states in the main
to the areas over which they initially asserted sovereignty have continued
significance? This approach can be summed up by the maxim *ex facto
oritur jus*. It does not however follow that the legal position evolving from
events is the same as that prior to those events. A contrary line to the last
approach is to treat the observance of the Treaty provisions by the
contracting parties with rights and claims, their failure to 'air and refresh'
at regular intervals these rights and claims before international forums,
and the omission of the non-recognising Contracting states to protest, as
effecting a 'loss of intensity' of the former legal position.[8] This erosion of
the former legal position for claimant, non-claimant and non-recognising
state alike may be welcome as providing the basis for the establishment
of new concepts and a new international regime in stage four of the
thinking.

Stage three

The immense achievements of a Treaty system which has survived
25 years now require coordination into the general system of international
law. Immediate points of stress have occurred in the interrelation of rights
of high seas and air space users under general law with the activities
permitted under the Treaty regime and within its area. For fish, the
Convention on Marine Living Resources has gone some way to resolve

them though it is too early to judge its success. For minerals there is a potential clash with the Deep Sea Bed regime provided in the Montego Bay Convention relating to the Law of the Sea, but in the light of the provisional understanding regarding Deep Sea Bed Mining 1984[9] (entered into by Belgium, France, West Germany, Italy, Japan, Netherlands, UK and USA), this may well undergo modification. There is no reason why, if and when such mining regimes become practicable, the two should not develop in parallel. As well as fish and minerals, Antarctica is an international source of fresh water, and has effects upon weather patterns throughout the world. This has led some to classify Antarctica as the common heritage of mankind along with the deep sea bed and outer space. Increased sophistication in analysis of the characteristics of these three areas highlights their difference. Antarctica is unique in having territorial claims as the fulcrum on which the regime was set in motion and there is a probability, at least for several decades, that its potential as a source of scientific knowledge is more likely to outweigh its commercial advantage, as compared to the deep sea bed and its military advantage, as compared to that of outer space.

Antarctica's potential as an international resource, however, raises considerations of sharing by the international community as a whole or by individual members not within the Treaty system and of participation of that community in the making of rules which govern the Antarctic regime. The rule-making authority for Antarctica is the Consultative states who operate by unanimity; the source of rules is the Treaty of 1959, Agreed Measures and Recommendations; less direct sources are the Antarctic Conventions adopted under the auspices of the Antarctic Consultative parties and the resolutions of the Scientific Conference on Antarctic Research (SCAR). Delegation of decision-making takes place through the appointment of observers and advance notice of expeditions and Antarctic stations, the grant of research contracts and permits, the designation of Special Protected Areas and Sites of Special Scientific Interest, the setting up of local arrangements such as the McMurdo Land Management and Conservation Board on which national agencies including the US Navy and Research Councils are represented. At present the system is institutionally immature, largely dependent on diplomatic negotiation for what is steadily developing into a complex network of relations. The need to systematise *ad hoc* arrangements and to provide adequate representation of global interests is likely to produce some increased institutionalisation in the decision-making procedures and clarification of the qualification and voting rights of Non-consultative and Observer states. Some delegation of the regulatory power of the Consultative Parties to an executive

Commission as foreshadowed in the arrangements in the Convention on Marine Living Resources may be necessary. Also the adoption of a Chamber system may ensure proper representation of non-recognising states alongside claimant and non-claimant states and at the same time preserve the executive control of the Consultative Parties.

Stage four

From the foregoing it is apparent that, if the successful impetus of the Antarctic Treaty regime is to continue, there must be a move away from the traditional territorial rights and claims of states, towards an objective regime. This regime should provide a workable balance according adequate participation to the international community in decision-making and in the rights and obligations of the Antarctic system and yet preserving the executive control of the group of states which have been most actively involved in the area. Such a regime will be achieved only by political negotiation. It will not be easy to do so as the degree of involvement in Antarctica of the full Consultative Parties differs and is disputed even among themselves. The international lawyer can aid the process by making available a number of institutional models drawn from the practice of regional and federal institutions and by spelling out the binding effect of such an objective regime on the position of third parties whether states, international organisations or individuals. In particular he should draw attention to the three interest groups, commercial, conservationist, recreational which at the present time advance positions for the commercial exploitation, preservation of the natural heritage (possibly as an international park), or utilisation for leisure or individual pursuit, which are not fully taken account of in government positions. (The strategic interest in the sense of maintaining world peace which may conflict with the defence interests of particular Treaty states is a possible fourth under-represented group.) There should be opportunity in the decision-making process for consultation with these interests either on a governmental or a non-governmental basis.[10] After decisions are made and implemented there should be some procedure by which they – and also the scientific interest currently represented by SCAR – may challenge the consequent activities as having a detrimental effect which unreasonably obstructs the pursuit of the particular interest in Antarctica.

The commercial interest is the most difficult to fit into this framework: yet, if excluded altogether, it is likely by external pressures and independent exploitation to bypass and defeat the present frail structure of Antarctic international cooperation. Coupled with a procedure of challenge might be adopted a presumption in favour of the *status quo* and the placing of the burden of proof on the developer to establish that this activity is of such a nature (as to location, process, transport etc.) as not unreasonably to disturb the balance or be detrimental to the other interests.[11]

The increase of activities in Antarctica must inevitably require the more frequent exercise of jurisdiction which at present by Article VIII of the Treaty is restricted to states' personal jurisdiction based on nationality. Issues of criminal and tortious responsibility and claims to property and rights arising out of contracts are likely to arise. It is for investigation whether the expansion of jurisdiction can be made in a novel manner and in one which might avoid some of the concurrent competing claims to jurisdiction endemic in present national systems of law. Various models have already been put forward and the current negotiations of a minerals regime have provided a spur to the thinking.[12] In the search for a model, the experience of states who have participated in joint development schemes for coal, oil, gas and other mineral resources (in some cases where sovereignty has been disputed) should be taken into account. One lesson appears to be that the establishment of a new system of law either by coordination of existing systems or a new code, involves much greater effort and uncertainty than the application of a single existing system. If one follows this lesson, the regime for an expanded system of jurisdiction in Antarctica might contain four features; first a narrow regulatory or criminal jurisdiction vested on a territorial basis in the Treaty states as a whole; second, rules drawn from private international law identifying the law for determination of the issue; third, a preliminary stage of compulsory arbitration for the reference of all disputes other than the serious reserved criminal offences; and fourth, some method whereby the special standards of conduct which Antarctic conditions impose can be taken into account in the determination of disputes.

As to the first, it might be possible to continue the present system whereby the state of the nationality of the wrongdoer has jurisdiction over homicide or serious assault. But it is likely that to preserve the aims of the Treaty and the balanced development of Antarctica the Treaty states will require a collective regulatory power to fine or punish serious offences. The list of such offences should be kept very short. As to the law applicable, it might be sufficient in all civil matters to apply the law of the state of the nationality of the defendant party to the dispute. The requirement of compulsory arbitration (whether with or without appeal or review to a national court) would help to reduce any nationalistic feeling in the dispute and to emphasise the Antarctic aspect of the case (which could be increased if the arbitrators were drawn from a panel of persons with knowledge of Antarctic matters). The introduction of Antarctic standards of conduct in the enquiry into criminal or delictual responsibility should also help to develop rules appropriate for application in Antarctica. For property and contract claims it might also be necessary to introduce a register of rights and claims and a special jurisdiction in connection with such a register might develop.

These are suggestions which involve far-ranging implications and much more detailed examination than is here possible. They are no more than essays to stimulate legal thought for the design of a legal system for Antarctica which will solve the conflicting claims of states and the consequential conflicts of jurisdiction arising from their territorial sovereignty.

At the outset I warned lawyers, other than the public international lawyer, that there was little of interest for them in the contemporary Antarctic scene. From my description of the four stages of legal thinking in relation to Antarctica this remains broadly true. However, when the fourth stage is reached, issues of private law, comparison of different national legal systems and the application of conflicts of laws rules will require the attention of all types of lawyers. At that stage it will be for the public international lawyer, working closely with these legal colleagues, to shape and operate a practicable legal system for Antarctica.

Endnotes

1 Application instituting Proceedings, *Antarctic Case (UK v. Argentina)*, 1955.
2 C. H. M. Waldock, 'Disputed sovereignty in the Falkland Island Dependencies' (1948) BYIL, **25** 311; O. Svarlien, 'The sector principle in law and practice' (1960) *Polar Record*, **10** 248; Auburn, *Antarctic Law and Politics* (1982), 17–31.
3 Waldock *op. cit.*, 317; Whiteman II, 1266; Auburn, *op. cit.*, 32.
4 J. Delbruck, 'Demilitarisation', 3 *Encyclopaedia of Public International Law* (1982) 150.
5 Draft Article IX and Article XIII (6).
6 UKTS No. 64, (1972) Cmnd 5000.
7 *Fisheries Jurisdiction Care (UK v. Iceland)* Merits ICJ Reports 1974, p. 3 at 112.
8 D. W. Greig 'Territorial sovereignty and the states of Antarctica' (1978) *Jo. Austr. Inst. of Int. Affairs*, **32** 117 at 128.
9 (1984) *ILM*, **23** 1354.
10 The USSR, unlike most other Consultative parties has opposed participation by environmental non-government organisations such as the Antarctic and South Ocean Coalition (ASOC) and the Union for Conservation of Nature and Natural Resources in the sessions of the Conference which drafted the 1980 Convention on Marine Living Resources. B. A. Boczek 'The Soviet Union and the Antarctic region' (1984) *AJIL*, **78** 834.
11 See Environmental Principles in the draft Convention on Mineral Resources, Article III.
12 D. W. Mouton, 'The international regime of the polar regions', *Hague Recueil*, **107** 175. R. Wolfrum, 'The use of Antarctic non-living resources: the search for a trustee'. In *Antarctic Challenge, Proceedings of an inter-disciplinary Symposium* (1983), 143.

8

The Antarctic Treaty System: some jurisdictional problems

GILLIAN D. TRIGGS

While seven states claim territorial sovereignty over all but a small wedge of the continent, other states, particularly those most active in the area, have refused to recognise these claims.[1] The difficulties raised by conflicting juridical positions on Antarctic sovereignty have persistently retarded negotiations concerning Antarctic resource and environmental regulation.[2] Indeed, the Antarctic Treaty and related Conventions and Recommendations are incomprehensible in the absence of some under-standing of the legal positions of claimant and non-claiming Parties on the question of territorial sovereignty.[3]

Despite the diametrically opposed arguments of states claiming terri-torial sovereignty in Antarctica and those denying it, and despite the complicating contentions of those states which deny existing claims but reserve their own rights to make claims in the future,[4] legal and diplo-matic techniques[5] have succeeded in avoiding direct conflict over sover-eignty. The debates in the First Committee of the General Assembly in 1984 demonstrate that most states, regardless of their ideological perspective, accept that the Antarctic Treaty regime has been a remarkably successful mechanism through which universal interests in preservation of the Antarctic environment, non-militarisation of the area, prohibition of nuclear explosions and radioactive waste disposal and free scientific access have been protected and advanced.[6]

The precise claims to sovereignty which are made and the legal principles upon which they are based have been amply and frequently described.[7] These claims are justly criticised as inappropriate principles for the determination today of territorial sovereignty in Antarctica.[8] The fact remains, however, that claimant states continue to parade a conglomerate of asserted classical theories in support of their territorial acquisitions – geographical contiguity, discovery, acquiescence, recognition, spheres of influence, effective occupation or manifestations of sovereign activities.[9] Legally out-dated or irrelevant as these devices may be, they form the juridical bases for existing claims to Antarctic sovereignty. For this reason, it is proposed to avoid repetition of the validity of, or defects in title, which arguably exist according to traditional legal principles. Rather, it is intended to discuss some of the legal difficulties which derive from the central conflict over sovereignty. These are first, the jurisdictional ambiguities which arise from the Antarctic Treaty and its relationship with developing customary and treaty law of the sea; second, the limited access to decision-making within the Antarctic legal regime; and third, the legal implications for Antarctica of the concept of a common heritage of mankind.

Jurisdictional ambiguities
Article IV
Critical to the success of the Antarctic Treaty has been the ingenious device of Article IV which is intended to preserve conflicting positions of claimant, potential claimant and non-claimant states, to deny legal effect to their activities during the life of the Treaty and to prohibit new or enlarged claims. While Marcoux aptly describes Article IV as creating a 'purgatory of ambiguity'[10] it facilitated ratification of an agreement which guarantees universal interests in Antarctic conservation and scientific research. The Article further provides a mechanism by which the issue of sovereignty is avoided in the subsequent interlinked Conventions on Seals[11] and Marine Living Resources.[12] Article IV thereby becomes binding on all the Parties to these further agreements. Similarly, the proposed minerals regime relies upon the link with Article IV of the Antarctic Treaty to avoid prejudicial implications for conflicting positions on sovereignty.[13]

It is doubtful, as a matter of treaty interpretation, that Article IV can have the legal effect of preserving the *status quo* of Parties prior to the Antarctic Treaty. This is partly because the Parties cannot be prevented from continuing to use their Antarctic bases in the event that the Treaty ends as a foundation for a sovereignty claim, and partly because the Treaty is not binding on third states and does not create an objective regime which is binding *erga omnes*.[14] It is more significant than the objective validity of Article IV, however, that the provision has been accepted by the Parties as a sufficiently secure foundation upon which to regulate activities in Antarctica and to conserve and study its environment. While the issue of sovereignty has clearly been a shadow over Consultative meetings and, indeed, it explains the restrained wording of some Recommendations,[15]

Article IV has had a positive function in enabling the Parties to move away from defensive domestic policies to respond to their common Antarctic interests.

Successful though this provision has been, it does not resolve legal difficulties over sovereignty which have arisen with the more recent developments of the law of the sea, both at customary international law and through the 1982 United Nations Convention on the Law of the Sea.[16]

Jurisdiction in coastal maritime zones

Where a state has sovereignty over littoral territory it *ipso facto* has sovereign rights over any adjacent continental shelf and territorial sea, whether or not a formal declaration or claim to such a shelf or sea has been made.[17] It logically follows that any state which claims sovereignty in Antarctica also has a claim to sovereign rights over its Antarctic continental shelf and territorial sea up to 12 nautical miles. A more difficult problem arises in relation to the exclusive economic zone (EEZ). The Convention on the Law of the Sea describes the sovereign rights which coastal states have over adjacent 200 mile exclusive economic zones; rights which now exist at customary international law and do not depend upon ratification of the Convention.[18] The question arises whether rights in the zone are inherent aspects of sovereignty which, like the continental shelf, do not need to be asserted independently. There are as yet no judicial pronouncements to resolve the issue. While exclusive economic zone and continental shelf rights are similarly described in the Law of the Sea Convention as 'sovereign rights' – thereby allowing the implication that they have similar juridical status – state practice has been formally to declare an exclusive economic zone.[19] For this reason, it is possible that an exclusive economic zone must be claimed and does not exist by virtue of the fact of territorial sovereignty.

This possibility is significant as any assertion of rights to an exclusive economic zone might conflict with Article IV(2), for it might constitute a new claim, or enlargement of an existing claim to territorial sovereignty. No such conflict arises in relation to the continental shelf, precisely because rights thereto are regarded as necessary incidents of territorial sovereignty. On the assumption that the express assertion of rights to an exclusive economic zone constitutes a claim to territorial sovereignty, such an assertion in Antarctica would be prohibited by Article IV(2). If so, the conclusion must then be that the freedoms of the high seas extend either to the limit of the exclusive economic zone if territorial sovereignty is recognised or, if it is not recognised, to the edge of the continental land-mass or iceshelf.[20]

It is submitted, however, that an assertion of rights over an exclusive economic zone is not contrary to Article IV(2). First, the better view is that an exclusive economic zone exists as a consequence of territorial sovereignty and does not depend upon prior assertion. Article 77(3) states that rights over the continental shelf 'do not depend upon occupation, effective or notional, or on any express proclamation'. While there is no such provision in relation to an exclusive economic zone, equally there is no such provision concerning the territorial sea, which is clearly established in international law. Both the *Beagle Channel*[21] arbitration and the *Aegean Sea Continental Shelf*[22] case support the view that the jurisdiction of a state over its adjacent maritime zone and continental shelf expands with the development of international law.

Second, a claim to an exclusive economic zone does not constitute a claim contrary to Article IV(2), for a claim to exercise sovereign rights in a maritime zone is not tantamount to a claim of *territorial* sovereignty.

State practice in Antarctica has been varied as to the declaration of exclusive economic zones. Argentina, Chile and France had, prior to negotiation of the 1982 Convention on the Law of the Sea, applied to their Antarctic territories their legislative claims to 200 mile exclusive economic zones.[23] By contrast, New Zealand[24] excluded the Ross Dependency from its declaration of a 200 mile exclusive economic zone in 1977 and similarly, Australia excluded the Australian Antarctic Territory from its 200 mile fishing zone declared in 1979.[25] Despite these exclusions, Australia and other claimant states continue to assert that they have a right to an exclusive coastal state jurisdiction over a 200 mile zone around their Antarctic territories.[26]

The draft Articles on an Antarctic minerals regime do not yet define the area which it is to encompass. At the Eleventh Meeting the Consultative Parties agreed that the regime should apply to 'all mineral resource activities taking place on the Antarctic continent and its adjacent offshore areas but without encroachment on the deep seabed'.[27] If the boundary were to be fixed at the 60th parallel South, like the Antarctic Treaty, the proposed regime would include continental shelves both of the Antarctic continent and of the South Sandwich Islands and the Heard and McDonald Islands which extend into the area. Thus some assessment of the risks of according jurisdiction to a minerals regime must be made by claimant states within the area.

The Antarctic Treaty applies to the entire area south of 60 °S latitude including the ice shelves south of that latitude. The ice shelf covering the land-mass is treated as tantamount to land and hence subject to a land regime.[28] It is a reasonable implication that the Treaty includes the high seas south of the 60 °S latitude, an implication which is supported by subsequent negotiation of the Convention on the Conservation of Antarctic Marine Living Resources in 1980.[29] This Convention covers the same area as the Antarctic Treaty but extends north of the 60 °S latitude to include the Antarctic Convergence.[30] It should be remembered, however, that the Antarctic regime thereby established is binding only on the ratifying Parties and not upon the rest of the international community. To the extent that the Antarctic Marine Living Resources regime establishes a rational system of management, and in light of rights of access to the Convention, there is every reason why Non-party states fishing or engaging in research in the area are likely to cooperate with the regime. As Professor R. Lagoni argues, the significance of the Convention lies in the principle of cooperation by which a treaty regime of rights and obligations supersedes the traditional freedom of access to living resources.[31]

Jurisdiction of the International Seabed Authority

The 1982 Convention on the Law of the Sea establishes a special international regime for the regulation of activities on the deep seabed described as the 'Area'. This is defined as

> Article (1) 'The seabed and ocean floor and subsoil thereof, beyond the limits of national jurisdiction.'

The words 'beyond the limits of national jurisdiction' raise the spectre of Antarctic sovereignty. If Antarctic claims are recognised then the jurisdictional limits of the International Seabed Authority will be the exclusive economic zones of the Antarctic claimant states. If Antarctic sovereignty is denied, some governments take the view that the Authority will have jurisdiction over the seabed in the area up to the shores of Antarctica (wherever they may be); thus including within the jurisdiction of the Authority the continental shelf zones most likely to be economically and technically feasible of exploitation in the future.[32]

Two criticisms are made of this view. First, it is clear that the Convention does not and was never intended to include Antarctica and its surrounding waters and deep seabed because the issue was deliberately excluded from the Conference as potentially divisive. Second, J. R. V. Prescott points out that the Convention does not envisage a situation where land does not include an exclusive economic zone or continental shelf other than rocks which 'cannot sustain habitation or economic life of their own', which were specifically excepted.[33] Both these criticisms support the conclusion that the International Seabed Authority would not have jurisdiction over the Antarctic continental shelf. This conclusion is supported by the political difficulties of claiming jurisdiction against the views of the Antarctic Treaty Parties, including those of non-claimant states such as the United States which has stated that the status of the continental shelf in Antarctica is 'unclear'.[34] Such a position is manifestly inconsistent with its view that no state has established territorial sovereignty in Antarctica. The Authority is also unlikely to pursue its potential jurisdictional control in Antarctica in favour of far more fruitful prospects of mining in other areas of the deep seabed.

While the threat of international control over the continental shelf under the Convention on the Law of the Sea is not significant for the foreseeable future, a literal interpretation of Article 1 of the Convention will undoubtedly provide support for those favouring the view that Antarctica is, in any event, the common heritage of mankind and subject to international jurisdiction.

Jurisdiction over Antarctic marine living resources

The claims by states to territorial sovereignty and to pertinent 200 mile maritime zones in Antarctica presented a significant obstacle to the creation of a single ecological regime for the entire area south of the Antarctic Convergence. There were two problems. First, as has been discussed, claimant states on the Antarctic continent assert a right to exclusive jurisdiction over the living resources up to 200 miles from the coast. Second, states whose sovereignty over islands such as Kerguelen and Crozet Islands situated north of the 60 °S latitude is unchallenged, but which have 200 mile exclusive economic zones within the Antarctic convergence, could justifiably argue that they too had exclusive jurisdiction within their maritime zones. Each assertion jeopardised the policy decision to establish a regime which considered the complex interaction of species of the unique Antarctic environment within a single ecosystem. It was decided to avoid the first problem by adopting a 'bifocal' approach.[35] This is a splendid example of contrived ambiguity by which all states can interpret the Convention in the way most suited to their respective juridical positions on sovereignty and thereby participate in the Convention. Article IV(2) of the Convention on the Conservation of Antarctic Marine Living Resources provides that

> Nothing in this convention and no acts or activities taking place while the present Convention is in force shall:
> (b) be interpreted as a renunciation or diminution by any Contracting Party of, or as prejudicing, any right of claim or basis of claim to exercise coastal State jurisdiction

under international law within the area to which this
Convention applies.

The words 'claim to exercise coastal State jurisdiction' allow claimant
states to interpret Article IV(2)(b) as referring to both undisputed islands
and disputed territorial claims north and south of the 60 °S latitude.
Non-claimant states may argue to the contrary that the right to exercise
coastal state jurisdiction exists only as to undisputed islands north of 60 °S
latitude.[36] The point is that, in this way, claimant states can participate
in the convention and protect their asserted sovereign rights through
Article IV(2)(b). While such a dual interpretation is of dubious legal
value,[37] its practical success depends upon the goodwill of all parties to
the Convention and, ultimately to its acceptance by the international
community. The obvious danger persists, however, that were a claimant
state to give effect to its view of Article IV of the Convention by, for
example, exercising the customary jurisdiction of a coastal state in matters
adjacent to its sectoral claim in Antarctica, the Convention would be
unlikely to survive.

The second difficulty was resolved by including in the Final Act of the
Conference a statement by the chairman to the effect that, where
sovereignty over an island within the area was recognised, the state could
decide whether the waters should be included in any specific conservation
measure adopted by the Commission. As the Commission cannot adopt
such a measure where the coastal state objects, the state's sovereign rights
can be preserved. While the Commission can adopt measures in the
claimed coastal waters of Antarctic claimants whose sovereignty is not
recognised, the Convention does not bind third states' activities within the
Antarctic area. Further, the Convention does not prevent claimant states
from enforcing a coastal state jurisdiction against their own nationals, as
is the case under Australian legislation,[38] nor against the nationals of
non-party third states. Again, this is because the Convention is limited in
legal effect to ratifying Parties.

The policy question for claimant states is whether to acknowledge
jurisdiction in the Commission over their claimed Antarctic coastal waters
implies a diminution of sovereignty. But, of course, Article IV is intended
to prevent any such implication. Article IV cannot, however, prevent third
states from drawing their own conclusions on this issue.

Limited participation in decision-making under the Antarctic Treaty Regime

The Secretary-General reported in 1984 that several states had criticised the Antarctic Treaty regime's decision-making processes which are restricted to the Consultative Parties.[39] In their view, the relatively free rights of accession are not adequate to compensate for the restriction of participation in decision-making to those 12 states named in the Preamble to the Antarctic Treaty and the 6 states (Poland, the Federal Republic of Germany, India, Brazil, China and Uruguay) which have subsequently demonstrated a capacity for substantial scientific research in Antarctica, thereby becoming Consultative Parties. The argument is that acceding states cannot afford the 'high entrance fee' required for Consultative Parties which effectively ensures the preservation of an exclusive regime to protect the interests of the privileged few.[40]

Other states deny the alleged exclusiveness of the Treaty structure, emphasising the wide spectrum of interests represented by the 20 states which have thus far acceded: industrial free market economies, socialist countries and developing countries.[41] At the Twelfth Consultative Meeting in 1983, these Non-consultative Parties were invited as observers and this was repeated at the Thirteenth Meeting. It is argued that, as all decisions are by consensus, the inability of the observers to vote becomes less important. While observers do not have a right to speak, they can play a role in the development of policy.

To the extent that the Treaty permits only limited access to the decision-making process, it can be justified on the ground that those states formulating measures for Antarctica should have a knowledge of and concern for the unique Antarctic environment.[42] There may be commendable sense in the view that a regional organisation of genuinely interested states is best able to manage Antarctic resources both in their own interests and in the interests of the international community. From the perspective of the approximately 130 states which are excluded from a decision-making role in Antarctica, such a rationale seems arrogant and self-serving. In light of these often emotive arguments, it might be useful to examine the Antarctic Treaty system to ascertain first, whether the allegations of an Antarctic 'club' are warranted and if so, second, what changes might be made to widen access to decision-making.

An examination of the Antarctic Treaty and linked Conventions and Recommendations confirms a preferred position for Consultative Parties. The Convention for the Conservation of Antarctic Seals is open for signature only to those states participating in the Conference which negotiated the Convention; the Consultative Parties to the Antarctic

Treaty.[43] Accession is open only to states invited to accede with the consent of the Contracting Parties to that Convention.[44]

The Convention on the Conservation of Antarctic Marine Living Resources 1980 is open for signature only to those states which participated in the negotiation for that Convention; the Consultative Parties to the Antarctic Treaty and the Federal Republic of Germany.[45] Accession is open to states 'interested' in research or harvesting activities in relation to marine living resources. Decision-making, however, is limited to the commission which comprises the Contracting Parties to the Convention, the Antarctic Treaty Consultative Parties and the Federal Republic of Germany, acceding states engaged in research or harvesting activities, and acceding regional economic integration organisations whose states' members are already participants.[46] A role in decision-making is thus dependent upon the qualification of 'interest' in research or harvesting. The term 'interest' is so vague as to be almost meaningless. Although it has been suggested that 'altruistic concern' alone may suffice,[47] it seems more likely that some practical involvement in research or harvesting will be required, thus possibly excluding many states from decision-making in the Commission.

The draft minerals regime does not describe the proposed Parties to it nor the rights of accession but, by necessary implication, it is clear that the Parties will include states which are not presently party to the Antarctic Treaty. As will be discussed in Part IV, however, the primary decision-making body, the Regulatory committee, is controlled by the existing Consultative Parties to the Antarctic Treaty.

In these ways, the Consultative Parties dominate the decision-making processes throughout the Antarctic Treaty regime. To emphasise the wide rights of accession available to Non-party states appears an empty consolation.

Parties acceding to the Conventions on Seals, Marine Living Resources and the proposed minerals regime also become bound by the Antarctic Treaty, the Agreed Measures and, where relevant, to the Conventions on Seals and Marine Living Resources. The Convention on the Conservation of Antarctic Marine Living Resources, for example, provides that any Contracting Party which is not a Party to the Antarctic Treaty must acknowledge the special obligations and responsibilities of the Antarctic Treaty Consultative Parties for the protection and conservation of the Antarctic Treaty area and to observe Recommendations of the Consultative Parties.[48] Through these linked obligations acceding states must accept the special role of the Consultative Parties and be bound by their measures both in the past and in the future.

While it is unclear what 'interested' means for the purposes of the Convention on Marine Living Resources, as Antarctic sealing takes place on a very restricted scale and Parties to the Convention on Seals play a preventive rather than regulatory role and as the structure of the minerals regime has yet to be agreed upon, the focus of the debate on participation in decision-making returns to the Antarctic Treaty itself. There is little doubt that acceding states will continue to demand Consultative Party status in the future even though they may not satisfy the objective requirements of substantial scientific research. It should be noticed that while such a requirement may involve a commitment of millions of dollars for states in no position to make it, there does not appear to be any corresponding obligation on the Consultative Parties to maintain an acceptable level of Antarctic research or activities. Indeed, the Secretary-General notes that 'not all [original Parties] have continuing research programmes in Antarctica at this time'.[49]

Article IX(2) sets a high standard for the demonstration of 'interest' in Antarctica with its examples of establishing a scientific station or despatching a scientific expedition. While it may be possible to interpret this provision in less rigorous or expensive ways, it is open to the Parties to amend the provision, perhaps by deleting the words after 'demonstrates its interest in Antarctica'. In this way the provision would be brought into line with the corresponding provision in the Convention on Antarctic Marine Living Resources (CCAMLR).[50] The difficulty with such a solution is that amendments to the Treaty require the unanimous agreement and subsequent ratification of all Consultative Parties.[51] An alternative method of amendment will come into existence in 1991 under which a Review Conference may be called where a majority of Contracting Parties, including a majority of Consultative Parties, may amend the Treaty.[52] If it is unlikely that the Consultative Parties will unanimously accept an amendment which gives decision-making power to acceding states which do not satisfy the requirement of 'substantial scientific research activities', it is more likely that such an amendment will be acceptable to a majority of Consultative Parties after 1991. For amendments to be achieved in this way, however, seems to be divisive and likely to threaten the future of the Antarctic regime. It would be more realistic and cooperative for Consultative Parties to recognise the significance of political and environmental interests of acceding states in Antarctic regulation, and to widen their access to decision-making as a unified policy well before 1991. The issue of participation in decision-making is thus a political rather than an essentially legal one.

The common heritage of mankind

As lawyers and diplomats engage in the delights of contrived ambiguities, bifocalism and restrictive interpretations of the Antarctic Treaty regime, political developments threaten to render their analyses otiose. Some states and commentators now argue that, as but one aspect of an ideological commitment to the New International Economic Order, Antarctica is, or ought to become, subject to the principle of a common heritage of mankind.[53] Various proposals for international control of Antarctica have been made in the 1940s by the United States[54] and in the 1950s by India,[55] New Zealand[56] and the United Kingdom.[57] However, the General Assembly Declaration of Principles Governing the Seabed and Ocean Floor and Subsoil thereof beyond the Limits of National Jurisdiction on 17 December 1970,[58] articulated the higher notion that resources of the area are the common heritage of mankind.

This notion has been accorded precise legal substance within the confines of the treaties on outer space[59] and the deep seabed.[60] It has been embraced as an attractive catch-phrase to embody the political ideal of equitable access to the earth's resources in the quite different context of Antarctica. The spirit of those arguing for the application of the principles of a common heritage to Antarctica is caught in a speech before the General Assembly by Amerasinghe, then the President of the United Nations Conference on the Law of the Sea, in which he said:

> There are still areas of this planet where opportunities remain for constructive and peaceful cooperation on the part of the international community for the common good of all rather than for the benefit of a few. Such an era is the Antarctic continent...there can be no doubt that there are vast possibilities for a new initiative that would rebound to the benefit of all mankind. Antarctica is an area where the now widely accepted ideas and concepts relating to international economic cooperation, with their special stress on the principle of equitable sharing of the world's resources, can find ample scope for application, given the cooperation and goodwill of those who have so far been active in the area.[61]

A lawyer's answer to this challenging concept of resource cooperation was presented by L. Ratiner, an administrator in the United States Ocean Mining Administration, in 1977. He argued that the

> political difference between the deep seabed and Antarctica and between the moon and Antarctica is stated quite simply –

territorial sovereignty, and a sovereignty claim, be it valid or dubious under international law, is nonetheless the grist of the international law mill.[62]

The conflict is thus between established legal principles of territorial sovereignty and more modern concepts of equitable sharing of the earth's resources. It arises in the context of Antarctic mineral resources not only because of an increased international concern for the preservation of the Antarctic environment, but also because the Antarctic Treaty Parties have embarked upon negotiations which at least envisage the possibility of mining the non-renewable resources. At present such mining is not economically viable or technically feasible, and the Consultative Parties have adopted a moratorium on mining.[63] Nonetheless, the spectre of Consultative Parties to the Antarctic Treaty regulating, apparently on their own behalf, mining in Antarctica and reaping any economic benefits which may accrue, was undoubtedly a galvanising factor for developing states to adopt by analogy the concept of a common heritage. The way in which the international community perceives the Consultative Parties' negotiations on minerals raises questions of international policy and strategy. First, however, it might be useful to examine the extent to which the principles of a common heritage have legal validity within Antarctica.

The concept of a common heritage for mankind is ill-defined and elusive. The term *res communis* to which the common heritage principle is likened, involves the notion of communal ownership and control of property which is, therefore, not subject to individual ownership or with sovereignty or sovereign rights of particular states.[64] While the notion of a common heritage is, at most, a principle rather than a set of rules, and thus does not entail specific and detailed legal requirements and consequences, it does have certain identifiable objectives. First, it rejects the notion of domination of property or of natural resources either by a particular state as sovereign, or by legal persons as 'owners' of property. Thus it is analogous to the notion of *res communis*. Second, it seeks to safeguard a common heritage for future use, and emphasises conservation of resources, in the sense of sustainable use and exploitation of those resources, and environmental protection. Third, the principle aims for an equitable allocation of the resources and benefits, with particular attention to the needs of developing States. Finally, it contemplates a legal regime to formulate precise rules to give effect to these general objectives.

It is ironic that the Outer Space and the Seabed Treaties took the Antarctic Treaty and its recognition of universal interests as precedents for their own more radical and precise articles.[65] Yet the states, both developed and developing, which negotiated the 1982 Convention on the Law of the Sea agreed to exclude Antarctica from consideration. This was primarily because it was believed that, while success was possible on the general issue of seabed resources, conflict over sovereignty was an insuperable obstacle to agreement on the question of Antarctic non-living resources. It is clear from an examination of the provisions of the Outer Space and Seabed Treaties that legal validity of the concept of a common heritage derives exclusively from these provisions and not from any developed customary law on the point. The result is that if the principle of a common heritage is to apply in Antarctica it must be because the treaty regime adopts them, specifically or in practice, or because customary international law has now embraced the principle in relation to all resources beyond national jurisdiction.

The Moon Treaty of 1979 provides that 'the moon and its natural resources are the common heritage of mankind' and establishes, when it becomes feasible, mechanisms to govern the exploitation of these resources.[66] Similarly, the 1982 Convention on the Law of the Sea provides that the 'Area and its resources are the common heritage of mankind' and that the resources are 'not subject to alienation' except in accordance with the rules, regulations and procedures to be established by the International Seabed Authority.[67] It has been significant in relation to both treaties that while adoption of the common heritage principle was acceptable to developed and developing countries, the specific provisions relating to the exploitation of resources have exposed the fundamental differences between these groups as to the precise obligations implicit in the notion.

The question whether unilateral exploitation of resources is necessarily incompatible with the wider interests of mankind has immediate relevance for the deep seabed miners which are nationals of states which are not, and do not intend to become, parties to the Convention on the Law of the Sea.[68] Certainly it is arguable that state practice supports the conclusion that a common heritage principle applies as a norm of customary law to deep seabed resources. It is, however, by no means clear whether such a customary norm specifically prohibits Non-treaty states from licensing their nationals to mine deep seabed resources outside the controls established by the Authority. Does respect for the common heritage require, for example, that exploitation should take place only under the control of a regime which operates for the benefit of all, especially developing states? Can states exploit 'common spaces' which are not presently under the jurisdiction of a management regime, if and when they have the capital and technology to do so?

It has been notable that the United States *Deep Seabed Hard Minerals*

Resources Act 1980 makes reference to its support for the common heritage principle 'with the exception that this principle would be legally defined under the terms of a comprehensive International Law of the Sea Treaty.[69] The implication is that the concept of a common heritage has legal substance only within the precise terms of the treaty which employs it and then, obviously, only in relation to those states which are party to it. This implication is supported by the earlier Moon Treaty which provides that the

> Moon and its natural resources are the common heritage of mankind which find expression in the provisions of this Agreement.[70]

While the notion of a common heritage has been invoked in relation to outer space resources, this has not resulted in the prohibition at customary law of unilateral exploitation by states with the necessary capital and technology. In relation to deep seabed resources, it is possible that a moratorium on unilateral exploitation now applies at customary international law, pending the establishment of an international regime.[71] But the status of deep seabed resources as *res communis* and the prohibition against their unilateral exploitation arises, not from the mere invocation of the notion of a common heritage, but from the fact that both propositions now amount, or will soon amount, to precepts of customary international law. They are particular propositions, applicable specifically to deep seabed resources, which have been refined during the nine years of negotiating the Convention on the Law of the Sea, and which have been accepted by most developed and developing States, including some states with the requisite technology and capital to exploit deep seabed resources and with a particular interest in so doing.

Thus the conclusion is justified that the principle of a common heritage does not invoke a legal phenomenon that is clearly understood at international law. Rather, it is a 'label for the bundle of provisions in the agreement creating a new type of territorial status'.[72] To employ the phrase a 'common heritage' in the absence of such a treaty, or readily identifiable customary rule, is to invoke a moral ideal or political aspiration which has no independent legal content. This conclusion has special relevance for Antarctica for if certain aspects of the common heritage principle apply there, unilateral exploitation of resources may nonetheless be permissible, and claims to territorial sovereignty may remain consistent with the notion.

There are certain aspects of the Antarctic Treaty regime and subsequent Consultative Party Recommendations and practices which might reasonably be interpreted as creating norms of international law, and hence rights and obligations applicable to Non-party states in Antarctica. In particular, these include obligations not to use the area for military purposes, to promote environmental preservation and to allow free scientific use. These obligations have arguably created expectations in the international community which should be given the status of law. The Antarctic Treaty prohibits all military measures including the establishment of military bases and fortifications, the carrying out of military manoeuvres or the testing of weapons. It establishes an inspection system to ensure compliance. Further, it establishes the first effective nuclear-weapons-free zone which bans nuclear explosions in Antarctica and the disposal of radioactive wastes. The Consultative Parties have adopted recommendations and agreements to protect the Antarctic environment and dependent ecosystems, including the Agreed Measures for the Conservation of Antarctic Fauna and Flora and the Convention for the Conservation of Antarctic Marine Living Resources. The Treaty guarantees freedom of scientific investigation and encourages international cooperation through the exchange of information, personnel and the results of scientific research.

It is very likely that these provisions and Recommendations now describe customary law for Antarctica which is binding on all states. The Preamble to the Antarctic Treaty recognises that 'it is in the interest of all mankind that Antarctica shall continue forever to be used exclusively for peaceful purposes and shall not become the scene or object of international discord'. Similarly, the Convention on the Conservation of Antarctic Marine Living Resources states that 'it is in the interest of all mankind to preserve the waters surrounding the Antarctic continent'[73] for the same purposes. Further, a Recommendation of the Ninth Consultative Meeting accepts that, in dealing with mineral resources in Antarctica, the Consultative Parties should not 'prejudice the interests of all mankind in Antarctica'.[74]

Some of these interests can be identified. They include Antarctica's influence of global weather patterns and its functions as a storehouse of 70% of the earth's fresh water.[75] Its unique and largely unspoiled environment make it a most significant scientific laboratory which could be kept intact if it were declared to be a world park. Any activities, and in particular mineral exploitation, which compromised these interests would violate customary norms relating to Antarctica.

Broadly, these norms may be described as protecting the interests of mankind in the Antarctic environment and as ensuring that certain benefits of use and exploitation accrue, or are available, to all states. The Antarctic

Treaty and related Recommendations and Conventions indicate that Antarctica is subject to the notion of a common heritage in these respects. However, such a conclusion does not describe the precise legal content which has been accorded the notion in the Convention on the Law of the Sea and the Outer Space Treaties. Does recognition of the interests of mankind in Antarctica imply the conclusion, for example, that territorial claims have, as a consequence, no validity; that exploitation of the living and non-living resources of the continental shelf of Antarctica are now subject to regulation by the International Sea Bed Authority under the Law of the Sea Convention; or that the economic benefits of resource exploitation must be distributed to developing states on an equitable basis? The short answer to these questions is that it does not.

It is too early to assert that the uniform and common usage of states, based upon *opinio juris*, is to regard Antarctica as *res communis*, and its resources as susceptible only to exploitation through an international regime. Such assertions by states which are not party to the Antarctic Treaty have been sporadic. Claimant states have persistently maintained their rights to sovereignty in the area and have generally acted consistently with those claims.[76]

Further, the analogy between the deep seabed and Antarctica is tenuous in other respects. Whether the inaccessible resources of the deep seabed were historically viewed as *res nullius* or *res communis* is immaterial. Since they have been recognised as part of the common heritage of mankind, it is likely that they should now be viewed as *res communis*. This status has been achieved before any state sought to establish effective occupation of, or territorial claims in relation to, portions of the deep seabed. Precisely the same is true of the resources of outer space. Indeed, even those states which have legislated unilaterally in relation to deep seabed resources deny that they intend to assert sovereignty over such resources.[77] It is not a radical step to argue that deep seabed and outer space resources should be managed by, and exploited through, an international authority, in the interests of the whole international community.

In marked contrast, Antarctica, prior to the intrusion of the notion of a common heritage of mankind, has been viewed as *terra nullius* by the international community and treated as such by states asserting territorial claims. Thus, 90% of Antarctica has been the subject of serious territorial claims for approximately 60 years. The prevailing situation between states, at the time when the notion of a common heritage illumined international consciousness, is thus significantly different from the deep seabed and outer space. It is unlikely that an international tribunal would refuse to acknowledge a consolidated title to territory in Antarctica, simply because the international community has expressed a demand that the resources are to be used in the interests of mankind.[78] This is particularly so in light of the persistent assertions and activities of claimant states with strong, though not exclusive, interests in the territory and its resources.

In conclusion, the legal consequences of invoking the notion of a common heritage in relation to the deep seabed and outer space do not presently apply in Antarctica. Although the notion of a common heritage has been invoked by some states in advocating a new regime for Antarctica, there is nothing to indicate that practical legal consequences which might flow from such a concept have, by state practice and *opinio juris*, achieved the status of principles of customary international law. If claims to territorial sovereignty in Antarctica have already been perfected, there is nothing in the notion of a common heritage which, as a matter of law, could supplant or displace those titles.

While the conclusion is that the principle of a common heritage does not yet amount to a rule of international law which denies validity to territorial claims to Antarctica and its resources, the principle possibly has the status of *lex ferenda* rather than *lex lata*.[79] It is a truism that if legal systems are to survive, they must be dynamic and flexible. Most especially, international law must address new international ideologies and recent concerns for Antarctica. It must attempt to develop beyond the traditional notions of territorial acquisition, which the Third World socialist states perceive as imperialist rationalisations for expansion. Certainly the future of Antarctica should not be jeopardised by capitulation to self-seeking ideologies, but it would be 'politically naive'[80] to assume that genuine concerns for Antarctic conservation and a share in its resources can be denied on the basis of outdated and inappropriate legal rules. For this reason, the conclusion that the principles of a common heritage do not deny sovereignty claims in Antarctica may fail to concede the evolutionary nature of law.

Conclusion

This chapter has explored some of the legal problems which continue to bedevil the Antarctic Treaty system and which are not fully resolved by the revered Article IV and its progeny. Questions remain concerning the jurisdiction of the International Seabed Authority and the jurisdictional limits of the Antarctic claimant states within their asserted maritime zones. The Convention on the Conservation of Antarctic Marine Living Resources puts such questions in a temporary abeyance which depends upon the fragile goodwill of the international community. Some moves for reform within the treaty system have been made. Certainly, opening Consultative Meetings to observation by acceding states is a small step in a realistic direction. Many more concessions of a less cosmetic kind will, nonetheless, be necessary to convince the Third World and socialist states that reform within the Antarctic Treaty system is likely to be more practicable and effective than under the auspices of the United Nations. The most important of such reforms include greater access to decision-making for acceding states and wider dissemination of scientific and environmental information. It will also be necessary to tackle the problem of jurisdiction over non-nationals for their criminal, contractual or tortious acts and for other matters such as employment and environmental regulation.

Any such reforms will, however, strike debilitating blows at the credibility of claims to territorial sovereignty. If the Antarctic Treaty system is to survive, these claims must be allowed to decline in importance and ultimately to wither away. Claimant states might usefully emphasise their genuine interests in Antarctica such as defence strategies, environmental protection and conservation, access to fisheries and minerals, and historical and cultural links. It may be that these interests can be safeguarded adequately within a reformed Antarctic Treaty system without dependence upon sovereignty claims which are highly unlikely to attract recognition within the international community. Indeed, while the phrases 'territorial sovereignty' or 'common heritage of mankind' give the debate about Antarctic resources an appearance of diametrically opposed positions, closer examination of the interests they embrace suggests that claimant states and the international community have much in common.

Endnotes

1 The seven claimants are the United Kingdom, Australia, New Zealand, Norway, France, Chile and Argentina. For the positions of these claims, see map at beginning of this volume. Note that about 15% of the continent remains unclaimed. The United States of America and the Union of Soviet Socialist Republics initially rejected the possibility of recognising any territorial claim in Antarctica. However, each state now reserves its right to make a claim in the future based upon discoveries and explorations by its scientists and explorers; Whiteman, *Digest of International Law*, 1248–50. There are also overlapping claims of Argentina, Chile and the United Kingdom which exacerbate the legal disadvantages of non-recognition; *Report of the Secretary-General to the Thirty-Ninth Session of the General Assembly*, 31 October 1984, A/39/583 (Part I) 17–20.

2 Difficulties over the jurisdiction of the Commission under the 1980 Convention for the Conservation of Antarctic Marine Living Resources are examples of this; see *infra*.

3 Note, for example, Articles IV and VIII of the Antarctic Treaty and Article IV(2) of the Conservation of Antarctic Marine Living Resources.

4 *Supra*, n. 2.
5 See discussion of this technique, *infra*.
6 General Debate, Consideration of and Action upon Draft Resolutions on Agenda Item 66 (Question of Antarctica), General Assembly, Thirty-Ninth Session, First Committee, 50th Meeting, 28 November 1984, A/C1./39/PV. 50.
7 Kish J., *The Law of International Spaces* (1973); Smedal G., *Acquisition of Sovereignty and the Status of Antarctica* (1931); Bernhardt J. P. A., 'Sovereignty in Antarctica' (1975) *Cal. W.I.L.J.*, 5, 297; Greig D. W., 'Territorial sovereignty and the status of Antarctica' (1978) *Australian Outlook*, 32, 117; O. Svarlien, 'The sector principle in law and practice' (1960–1) *Polar Record*, 10, 248; G. Triggs, *International Law and Australia's Sovereignty in Antarctica* (1986) Legal Books, Sydney; 'Thaw in International Law, Rights in Antarctica under the Law of Common Spaces' (1978) *Yale L. J.*, 87, 804.
8 For these criticisms see authors cited *ibid*, especially Bernhardt and Greig.
9 For a discussion of general principles see Y. Z. Blum, *Historic Titles in International Law* (1965); R. Y. Jennings, *The Acquisition of Territory in International Law* (1963); I. Brownlie, *Principles of Public International Law*, (3rd edn, 1969); D. P. O'Connell, *International Law* (2nd edn, 1970) Vol. I, 403ff.
10 Marcoux, J. M. 'Natural resource jurisdiction on the Antarctic continental margin' (1971) *V.J.I.L.*, 11, 374, 379.
11 Convention for the Conservation of Antarctic Seals, 1972 UKTS No. 45 (1978) (Cmnd 7209).
12 Convention on the Conservation of Antarctic Marine Living Resources 1980, *Polar Record*, 20, 385–95; J. N. Barnes, 'The emerging Antarctic Living Resources Convention' (1970) *A.S.I.L. Proc.*, 272.
13 See the draft proposals on a minerals regime known as 'Beeby I' published by the Friends of the Earth (1983) *ECO*, 23(1), 5; for a discussion of these proposals see G. Triggs (1984) *Australian Mining and Petroleum Law Association Yearbook*, 525.
14 There is debate about the question whether an objective regime is created by the Antarctic Treaty; see, for example, the Soviet view that the Treaty is valid *erga omnes*, *Kurs Mezhdunarodnovo Prava* (A Course of International Law), Vol. 3, 402; Waldock, C. H. M. (1964) *Y.I.L.C.*, 2, 29, 30; cf. F. M. Auburn, *Antarctic Law and Politics* (1981) 117–18.
15 Note, for example, the weak provisions concerning implementation of the Agreed Measures for the Conservation of Antarctic Fauna and Flora 1964, (1965) *Polar Record*, 12, 457–62, Article III.
16 Although the Convention on the Law of the Sea is not yet in force it attracted 130 votes in the General Assembly (with 4 against and 17 abstentions) and articulates and describes many changes in customary international law: UN Doc. A/Conf. 62/122; *I.L.M.*, 21, 1201 (1982).
17 Convention on the Law of the Sea 1982, Arts 3, 76, 77.
18 Articles 55–75; D. P. O'Connell, *The International Law of the Sea* (1982) Vol. I, 570; by 1980, 95 states had claimed 200 mile jurisdictional limits.
19 O'Connell, ibid., 570–1; note that some declarations are limited to fisheries regulation, while others claim full 200 mile territorial seas.
20 Yet another legal issue concerns the precise delimitation of boundaries as the iceshelf, which is for practical purposes permanent, may extend many miles beyond the continental land-mass which the shelf depresses, D. Pharand, *The Law of the Sea of the Arctic: with Special Reference to Canada* (1973) 181.
21 (1978) *I.L.M.*, 17, 634.
22 (1978) ICJ, Reps 1.

23 Auburn, F. M. *op. cit.*, n.15, 219; W. M. Bush, *Antarctica and International Law* (1982), Argentina 63, Chile 447, France 584.
24 *Territorial Sea and Exclusive Economic Zone Act*, 1977, s. 2.
25 On 20 September 1979 'Australian Waters' under the *Fisheries Act* 1952 were proclaimed to be 200 miles from the baselines. On 31 October 1979 the waters of the Australian Antarctic Territory were excluded, Proclamation, 31 October 1979, *Commonwealth of Australia Gazette* N.S. 225, 1; *Fisheries Amendment Act* 1980 (no. 86 of 1980). Note that the *Fisheries Act* 1952 continues to apply to the activities of Australian citizens fishing in waters abutting the Australian Antarctic Territory.
26 For Australia's position see 'Antarctica – a continent of international harmony' (1980) *Aust. Foreign Affairs Record*, **51**(2), 10.
27 Recommendation XI(iv).
28 Pharand, D. *op. cit.* n. 21, 181; G. D. Triggs, *op. cit.*, n. 8.
29 The Marine Fishing Resources Convention applies within the Antarctic convergence defined in Article I(4).
30 This, of course, assumes that the regime is not binding *erga omnes*.
31 M. Koch, 'The Antarctic challenge: conflicting interests, co-operation, environmental protection and economic development' (1984) *J. Maritime Law & Comm.*, **15**(1), 117, 121.
32 *Secretary-General's Report, op. cit.*, n. 1, 27.
33 Prescott, J. R. V., 'Boundaries in Antarctica', Harris (ed.) *Australia's Antarctic Policy Options* (1984), 110.
34 'U.S. Antarctic Policy' Hearing before the Subcommittee on Oceans and International Environment of the Committee on Foreign Relations, U.S. Senate, 94th Cong. 1st. Sess. (1975) 68.
35 For an explanation of this technique see, *Sec.-Gen's Report, op. cit.*, n. 1, 61–2; Barnes, J. N. *op. cit.*, n. 13, 264–94, 276.
36 R. Zagoni, 'Convention on the conservation of Marine Living Resources: a model for the use of a common good?', *Proceedings of an Interdisciplinary Symposium*, 97.
37 It is at least possible that an international tribunal might be required to consider the objective legal meaning of Art. IV(a)(b). Principles of treaty interpretation would then be decisive of the question as to its meaning, although state practice subsequent to a treaty can have bearing on the issue.
38 *Supra*, n. 26.
39 *Sec.-Gen.'s Report. op. cit.*, n. 1, 35.
40 Ibid., 29.
41 See, for example, the views of Mr Richard Woolcott, Australian Ambassador to the United Nations, at the Workshop on the Antarctic Treaty System, Beardmore Glacier, Antarctica, 5–13 January 1985 *Antarctica*, 8.
42 Ibid.
43 Article 10.
44 Article 12.
45 Article XXVI.
46 Article VII(2).
47 Auburn, F. M. *op. cit.*, n. 15, 237.
48 Article V(1).
49 *Sec.-Gen.'s Report, op. cit.*, n. 1, 29.
50 Article VII of the CCAMLR permits acceding States to become members of the decision-making Commission 'during such time as the acceding party is engaged in research or harvesting activities in relation to the marine living resources to which this Convention applies'.

108   *Legal issues*

51 Antarctic Treaty, Article XII(1).
52 Ibid., Article XII(2).
53 'Thaw in International Law?' *op. cit.*, n. 8, 821; Bernhardt, *op. cit.*, n. 8, 349;
   M. C. W. Pinto, 'The international community and Antarctica' (1978)
   U. Miami L.R. **33**, 472, 478–9; Kish, J. *op. cit.*, n. 8; P. Jessup and
   H. J. Taubenfeld, *Controls for Outer Space and the Antarctic Analogy* (1959).
   See the *Sec. Gen.'s Report*, *op. cit.*, n. 1, Part 2, for the views of 54 states
   responding to a request for information in accordance with Gen. Ass. Res.
   38/77.
54 Hayton, 'The Antarctic settlement of 1959' (1960) *A.J.I.R.*, **54**, 349.
55 India requested the UN General Assembly to consider the 'Question of
   Antarctica' on the ground that its size and international importance rendered it
   'appropriate and timely for all nations to agree and to affirm that the area will
   be utilized entirely…for the general welfare', UN Doc. A/3118 (1956).
56 Hannessian, J. 'The Antarctic Treaty 1959' (1960) *I.C.L.Q.*, **9**, 436, 452–5.
57 Ibid., 436, n. 5; W. Nash, Prime Minister of New Zealand proposed in 1956
   that Antarctica be established as a 'world territory under UN control'.
58 GA Res. 2749 (XXV), *I.L.M.* **10**, 230 (1970), adopted by 104 votes to 0, with
   14 abstentions.
59 Treaty on Principles Governing the Activities of States in the Exploration and
   use of Outer Space, including the Moon and other Celestial Bodies 1967,
   U.K.T.S. 10 (1968) Cmnd. 3519; 610 U.N.T.S. 205; in force October 10, 1967
   with 84 contracting parties; 1979 Agreement Governing the Activities of States
   on the Moon and Other Celestial Bodies (1979) 18 I.L.M. 1434, not yet in
   force, for a discussion of this Agreement see (1981), 74 *Proc. A.S.I.L.*, 152–74.
60 *Supra*, n. 17.
61 Secretary-General's Report, *op. cit.*, n. 1, 64.
62 Leigh Ratimer, Statement, 'Earthscan', Press Briefing Seminar on the Future
   of Antarctica, London, 27 July 1977.
63 Recommendation of the Ninth Consultative Meeting, Rec. IX 1(8). The
   Consultative Parties agreed to

   > exploration and exploitation of Antarctic mineral resources while
   > making progress towards the timely adoption of an agreed regime
   > concerning Antarctic mineral resource activities. They will thus
   > endeavour to insure that, pending the timely adoption of agreed
   > solutions pertaining to exploration and exploitation of mineral
   > resources, no activity shall be conducted to explore or exploit such
   > resources.

64 O'Connell, D. P. *op. cit.*, n. 10, 406; Brownlie, I. *op. cit.*, n. 10, 181; E. Borgese
   (ed.) *Pacem in Maribus* (1972) 161.
65 Note, in particular, the Preamble to the Antarctic Treaty.
66 *Supra*, n. 83, Article 11(1).
67 *Supra*, n. 17, Articles 136 and 137.
68 This issue concerns the United States, United Kingdom, Federal Republic of
   Germany, Japan, Soviet Union, France and Italy. Note, however, that only the
   United States is firmly committed to unilateral mining; Oxman, B. H., Caran,
   D. D., Bueri, C. L. O., *Law of the Sea: U.S. Policy Dilemma* (1983).
69 94 Stat. 553, Sec. 2(a)(3), (1980) *I.L.M.*, **19**, 1003; (1981) *I.L.M.*, **20**, 1228.
70 Article 11(1), see also, B. Cheng, 'The Moon Treaty: agreement governing the
   activities of states on the moon and other celestial bodies within the solar
   system other than the earth', 18 December 1979 (1980) *Current Legal
   Problems*, 213, 222.

71 Reasonable, and less than reasonable, legal minds differ markedly on this question. See a presentation of negotiating positions of the United States and the Group of 77 extracted in Oxman, B. M. *The Third UN Conference on the Law of the Sea: Seventh Session 1978* (1979) *A.J.I.L.* **73**, 1, 35, n. 119.

72 Cheng, *op. cit.*, n. 70, 222.

73 This, of course, depends upon whether these imputations have 'gelled' into rules of custom, O'Connell, D. P. *op. cit.*, n. 10, 15; Akehurst, M. 'Custom as a source of international law' (1974–5) 47 *B.Y.I.L.*, **47**, 1, 2; J. Finnis, 'Theory and methodology of international law' Report of the 58th Conference of the I.L.A., Manila, 1978, 186–206.

74 Preamble.

75 Rec. IX-1 (4)(iv); see also, Rec. IX-5, IX-6.

76 Elliott, H. *A Framework for Assessing Environmental Impacts of Possible Antarctic Mineral Development* (1977) Institute of Polar Studies, Ohio State Univ., II-1.

77 For translation of those available documents in which each of the claimants maintains its claim see Bush, W. M. *op. cit.*, n. 24, Vols I and II.

78 France, the FRG, Japan, the UK and the US have enacted unilateral legislation which purports to accept the principle of a common heritage and which set up trust funds for a percentage of the benefits to be distributed by the Seabed Authority once it comes into existence. See, for example, UK Deep Sea Mining (Temporary Provisions) Act 1981, s. 10.

79 A tribunal is likely to resort, on the first instance, to the classical rules for the acquisition of territory. It may, however, consider international interests in Antarctica now and under precedent inappropriate; see Greig, D. W. *op. cit.*, n. 1, and his commentary in Harris, S. *op. cit.*, n. 34, 75; see also, G. D. Triggs, in Harris, S. *op. cit.*, n. 34, 29–66.

80 Woolcott, R. *op. cit.*, n. 41, 12.

Part III

The Antarctic Treaty Regime: protecting the marine environment

9

Introduction

Since the 1970s the Consultative Parties have been concerned to regulate and to minimise man's impact on the Antarctic environment. At the Ninth Consultative Meeting the Parties formally recognised 'their prime responsibilities for the protection of the Antarctic environment from all forms of human interference'. They have since recommended a Code of Conduct for Antarctic expeditions and station activities, established a meeting of experts to study the effects of oil pollution on the Antarctic environment, adopted a Statement of Accepted Principles and Good Conduct Guide for Tourist Groups and recommended that environmental impact studies be made to evaluate major operations proposed in the Antarctic Treaty area. These measures are hortatory not mandatory, and they depend for their effect upon implementation by those states which choose to accept them.

In 1980 the Consultative Parties negotiated the Convention for the Conservation of Antarctic Marine Living Resources. This Convention is remarkable in many respects. Firstly it was negotiated, ratified and came into force with great speed; a speed which was fuelled by the urgent need to establish some form of regulation for the harvesting of krill and other Antarctic fisheries. The Convention is notable, secondly, for its adoption of a single ecosystem approach to conservation in which the Antarctic Convergence provides the outer limits of jurisdiction. Thirdly, and for the first time within the Antarctic Treaty system, a permanent structure was established by the Convention to give effect to its aims and objectives. The Convention establishes a Commission which includes all original contracting parties and those acceding states which are engaged in Antarctic marine research or harvesting. Its functions are, *inter alia*, to adopt conservation measures (on the advice of a Scientific Committee) and to implement a system of observation and inspection. Conservation measures may include the designation of the quantity of species which may be harvested, of their size, age and sex, of open and closed harvesting seasons and of protected species. While decisions of the Commission on matters of substance are by consensus, they become binding upon members of the Commission within 180 days of notification, unless a member advises the Commission that it is unable to accept the measure. In this event, the measure is not binding upon that member. The Consultative Parties have thus achieved little more than a voluntary code. However, the political will to demonstrate that the Antarctic Treaty system is responsible and effective has been strong and, for this reason, it is very likely that members will act in accord with any measures adopted by the Commission.

Not so remarkable was the adoption of another variation on the theme of Article IV in order to avoid the problem of conflicting juridical positions on claims to territorial sovereignity in Antarctica, and various interlinking provisions by which the Parties become bound by obligations of the Antarctic Treaty itself and of the Agreed Measures and other Recommendations made under the Treaty. It is through these interlinking obligations that a more comprehensive structure for the regulation of Antarctic resources is being evolved. It is also notable in this regard that all parties to the Antarctic Treaty are automatically members of the Commission, and these members will be joined by such additional states as accede to the Convention. Any acceding party is entitled to be a member of the Commission only for as long as it is engaged in research or harvesting activities in relation to Antarctic marine living resources. In this way, the Antarctic Treaty Consultative Parties retain control over the regulation of marine living resources within the Antarctic Convergence.

The Convention entered into force on 7 April 1982 and in that year the first full meeting of the Commission was held at its headquarters established in Hobart, Australia. Meetings have since been held annually. At present, Argentina, Australia, Belgium, Chile, the European Community, France, the Federal Republic of Germany, the German Democratic Republic, Japan, New Zealand, Norway, Poland, South Africa, the United Kingdom, the United States and the Soviet Union are full members of the Commission. Spain and Sweden have acceded to the Convention without joining the Commission.

The Commission has been concerned in its early meetings with essentially legal, administrative and financial aspects, however, at its meeting in September 1984, the Commission made significant moves towards the conservation and management of Antarctic marine living resources. Emphasis lay upon the depleted stocks of fin fish in the Antarctic, especially in the South Atlantic sector where an annual catch of fin fish is about 120000 tonnes, mostly taken by the Soviet Union. The Commission agreed to binding conservation measures banning all fishing within 12 nautical miles of the island of South Georgia, (a major spawning ground for *Notothenia rosii*, a species which has suffered a severe depletion in recent years). Measures have also been adopted to protect juvenile fin fish throughout Antarctic waters through limits on net mesh sizes. Members have also agreed to take efforts to otherwise limit the size of fish which are caught.

At the meeting in 1985, Parties recognised that a major problem facing CCAMLR is the lack of data and research. It is with this in mind that the Commission has asked members to begin programmes of research to establish what further measures are necessary to assist the recovery of stocks of depleted fish. It is ironic that, while the Convention was negotiated to protect krill, annual catches of this crustacean have fallen from about a half million tonnes in 1981–2, to about a quarter million tonnes in 1985; a level which is considered by the scientific committee as unlikely to deplete stocks. While it is too early to make any serious assessment of the success of CCAMLR, it is very likely that the credibility of the Antarctic Treaty Consultative Parties to manage Antarctic resources will be seen to rest with its ability to adopt and enforce conservation measures within the Antarctic convergence.

Discussion at the British Institute's conference considered the most effective means to enforce the conservation of living resources. Catch limits, whether set by international agreement or by coastal state legislation, have, in the past, been too generous or ineffectively enforced to control commercial exploitation of a fish stock to a level beyond natural replenishment. Would the Antarctic Treaty system be any better able to manage these stocks? Some form of management of stocks was advocated in addition to conservation measures. It was noted that recommendations made by SCAR and other scientific bodies were not always taken up by the Consultative Parties. However, increased weight to such recommendations might now be given in the light of recent initiatives to achieve closer cooperation between SCAR and the International Union for the Conservation of Nature (IUCN) and now that SCAR has observer status in CCAMLR.

Discussion also considered the second stage of processing data to produce the information upon which to set and to update catch limits. It was believed that an assumption of power by Consultative Parties to manage the Antarctic environment carries with it the responsibility to facilitate access by Third World states and their nationals to Antarctica. Finally, some delegates stressed that mineral exploitation on a limited and properly regulated basis would have very little environmental effect upon so large an area as the Antarctic continent.

10

The Antarctic Treaty system as a resource management mechanism

J.A.GULLAND

Introduction

In the Antarctic the living resources offer a complete contrast between marine and terrestial systems. The Southern Ocean is rich with life – among which krill, whales, seals and penguins are the best known; historically, man's visits to the Antarctic have been largely those of sealers, whalers and fishermen engaged in harvesting these resources. The management problems are those of ensuring that this harvesting is carried out in a rational manner, with due regard to future interests in the resources. The Antarctic land mass is cold and barren and extremely hostile to life. What life there is, is probably vulnerable to disturbance through man's other activities, such as research, tourism, or, possibly in the future, extraction of minerals. The management problem is that of diminishing or minimising that disturbance.

The management problems of sea and land are therefore best discussed separately. The exception is the needs of seals and penguins for a firm base (land, or ice-shelves) on which to breed. These essentially marine animals can be vulnerable to damage to, or disturbance at, these breeding sites, and mechanisms to prevent this are best discussed together with other aspects of terrestrial management.

Marine resources

The background

While the Antarctic Treaty applies to the seas south of 60° S, in considering the marine resources it is better to look at the whole area south of the Antarctic Convergence. The exact position of the Convergence is variable, but on average corresponds closely to the boundary of the area of responsibility of the Commission for the Conservation of Antarctic Marine Living Resources (CCAMLR). South of the Convergence the

116

current systems ensure a good supply of nutrients to the surface and, in the summer, the primary production in parts of the region in the form of microscopic plants (phytoplankton) is among the highest in the world, with the exception of a few spectacular areas such as the upwelling zones of Peru and California.

A further favourable factor from man's point of view is that the food chain is generally short. There are, for example, two steps from phytoplankton to baleen whales via krill. This means that a relatively high proportion of the original primary production appears in a harvestable form, and the standing stock represents the accumulated production of several years (several decades in the case of whales), and is therefore high.

The waters of the Antarctic and sub-Antarctic, despite being cold, rough, and far from civilisation, have attracted sealers, whalers and fishermen for the past two centuries. The use of these resources shows that unmanaged exploitation can be disastrous. By the end of the nineteenth century, the fur and elephant seals and right whales of the sub-Antarctic had been brought close to extinction; by the middle of this century the larger baleen whales (blue, fin and humpback) had also been greatly reduced. Partly through the efforts of the International Whaling Commission (IWC), their depletion has been slightly less extreme than that of right whales or fur seals, but of the baleen whales only the minke whale is now present in similar numbers to those at the time man first came to the Southern Ocean. Exploitation is now focused on krill and demersal (bottom-living) fish. Several fish stocks seem already, as reported at the recent meeting of CCAMLR, to be greatly reduced from their original level. Krill are so far probably little affected. Current catches, at around half a million tons, are still far less than the potential yield. Estimates of this range from tens of millions of tons upwards, i.e. comparable to the present total yield of all types of fish from the oceans of the world.

The declines or collapses of seals, whales and fish are not unique to the Southern Ocean. Virtually every commercially attractive exploited fish stock has been allowed to decline below its most productive level. These declines have not often turned into catastrophic collapses largely because of two factors – the high reproductive capacity of most fish makes them less vulnerable to sustained over-exploitation than mammals, and most fishing fleets can move to alternative resources when stocks and catch rates in one area begin to decline. The events in the Southern Ocean should therefore not be ascribed to some unusual degree of greed or short-sightedness on the part of the harvesters, but are the predictable results of unmanaged exploitation of a common-property, open-access resource.

From the point of view of managing the marine resources of the Antarctic these points can be made.

i) The resources are rich, and have made (and could continue to make) significant contribution to world food supplies.

ii) Several important elements (fur and elephant seals, and whales) of the marine ecosystem have already been greatly altered, and this has probably had effects on many other elements, i.e. in the ocean we are not dealing, as is the case for the Antarctic land mass, with an undisturbed system.

iii) If the resources are to be maintained in an optimum condition management will be essential.

The following sections of this chapter will discuss the mechanisms required to provide the necessary management actions (including deciding upon what is meant by optimum condition in (iii) above), and then, with special reference to the purpose of this volume, examine the past, present, and potential future role of the Antarctic Treaty in ensuring that these mechanisms are established and operate correctly.

Mechanisms for management

The basic requirements for successful fishery management have been discussed on a number of occasions, notably by the Food and Agriculture Organization's (FAO) Advisory Committee on Marine Resources Research and its working parties. The need is not merely to have a mechanism to introduce management measures. Before such action and any decision on specific measures, it is necessary to have discussions and agreement on the objectives which the measures should achieve, and adequate technical and scientific information on the immediate and long-term results of alternative management actions (including the possibility of doing nothing). The actions must then be followed by steps to ensure that the agreed measures are actually implemented, and to review their success and where necessary to revise them.

At each stage it is important that all those actually or potentially interested in the resource should participate. Without virtually unanimous agreement few measures are likely to be effective. Any participant that does not abide by these (e.g. in limiting the total volume of his catches, or by avoiding catching small or immature animals) will gain nearly all the benefits of the management actions of others, and there will be little net conservation effect. Bitter experience has also shown that management measures as well as being unanimous should also be introduced as early as possible, before the industry builds up excess capacity. The aim should be to do little more than put the brakes on development as the optimum harvesting rate is approached, rather than having to cut back on over-capacity, with all the economic and social problems this is likely to bring. Unfortunately, the first criterion works against this. All the participants are only likely to agree on the need to do something once the need has become incontrovertible with the collapse or severe depletion of the resource.

How do events in the Antarctic fit in with these requirements? The fur seal and right whale stocks collapsed before there was any attempt at management, and few lessons can be learnt. Much more can be learnt from the IWC, which suffered from several serious flaws, though it must not be assumed that everything the IWC has done was wrong. Indeed, when first established in 1946, it could be considered as being in the forefront of what would later become the environmental movement.

The first flaw to become apparent was the lack of good quantitative information on what was happening to the stocks, and what, in quantitative terms, would be the effect of different management measures (in particular the effect of a reduction in the quota). Only when fishery scientists, who had taken a more quantitative approach, were brought in, as members of the IWC's Committee of Three (later Four) Scientists to advise the Commission in the early 1960s, did the Commission have adequate information on which to base its policies. The blame for this – if it is fair to talk about blame with the benefit of many years of later experience – lies further back with those that determined the level and direction of research into the Southern Oceans. It was particularly unfortunate that after the UK government had established the Discovery Committee and launched a major research effort (and which was set up, in part, explicitly in support of the whaling industry), no-one ensured that quantitative studies of the dynamics of whale stocks, and their reaction to exploitation, were actually carried out.

Other flaws became apparent after the Committee of Four had made their report. It became clear that great reductions in quotas were necessary and that problems existed concerning the objectives of the Commission, and participation in its deliberations. After the Commission received, from 1964 onwards, clear scientific information on what needed doing, it had, until the mid 1970s, great difficulty and considerable delays, in acting on it. This was because of objectives in the original Convention which implied, or could be interpreted as implying, that considerable weight should be given to the economic interests of the whaling industry. This problem was compounded by the fact that membership of the Commission was confined largely to countries with active whaling industries, whose immediate economic interests were a major factor in determining national policies.

The situation is now reversed, with many of the countries having no direct connection with whaling, and the policies of some former whaling countries being largely determined by active environmental lobbies. The result is that decisions of the Commission tend to be as strongly weighted against any harvesting of whales as they previously were in favour of a volume of catches in excess of what the resource could sustain. There is, for example, as little scientific justification (assuming the objective of the Commission is to ensure rational use of the resource) for the currently recommended moratorium on all minke whale catching as there was for the continuation of a total quota of 15 000–16 000 Blue Whale Units around 1960.

Events in the utilisation and management of other resources (krill, fish and the seals of the ice-shelf) have not advanced significantly to make a judgment of the management arrangements. Special mention should however be made of the 1972 Conference on the Conservation of Antarctic Seals. This conference was called on the initiative of the Antarctic Treaty powers, at a time when there was no commercial harvesting of seals, but when there was a possibility that such sealing might start. Since it recognised that commercial sealing was an intrinsically legitimate activity, if done in a responsible manner, it has sometimes been regarded as a pro-sealing conference and convention. This is very far from being true. Those responsible for calling the conference were well aware of the dangers of uncontrolled harvesting, and the need to set up proper controls early. For about the first time in history the necessary framework to institute control was set up in advance of any commercial activity. The Convention gives the interested countries the rights and responsibilities to manage any sealing, and, in the annex, spells out specific controls (annual catch limits, and a pattern of closed areas), which should ensure that any harvesting is well within the productive capacity of the stocks, pending more precise scientific analysis.

The more recent convention, establishing CCAMLR, was set up while the krill fishery were still growing, and while the catches were well below the likely level of the sustainable yield, and before there was any evidence that the fishery was affecting the stock. The fishery on demersal fish around several of the Antarctic and sub-Antarctic islands had reduced some of the stocks well below their unfished level, but there are some biological reasons to be less seriously concerned about such reductions in fish stocks than similar reductions in mammals. The real concern which led to the establishment of CCAMLR was less for krill or fish in themselves, than for the impact that a large-scale krill fishery might have on those species that feed on krill. Krill plays a central part in the Southern Ocean ecosystem, being the most abundant herbivore. It is the major food of many of the larger animals, including baleen whales, penguins, and crab-eater seals. There is a fear that if the krill abundance were reduced by fishing, this could endanger the recovery of the depleted stocks of baleen whales.

This probable interaction between krill fishing and the dynamics of whale population illustrates two points that are becoming increasingly apparent in present-day resource management: that the objectives of management are complex, and that in order to achieve whatever objective is decided upon, quite detailed scientific research is likely to be needed.

If the Southern Ocean were managed purely in order to maximise the supply of food from the region it is probable that harvesting should be concentrated on krill (assuming the technological and economic problems of catching, processing and marketing of huge quantities of krill can be solved) with the whale and seal stocks being allowed to decline. Against this, there are those who feel that nothing should be allowed that would prevent a rapid recovery of whale stocks to their original unexploited level. A more balanced and more generally acceptable policy might be for a krill harvest rather less than the maximum possible, sufficiently small to offer no threat to the survival of the whale populations, though probably not so small as to permit any harvesting of whales.

Whatever objective is chosen it is clear that those setting quotas or other appropriate measures for krill will need to know what the long-term effect is on stocks of krill, and on those of whales and other consumers, of different patterns of krill harvest. It is probable that the answer will depend not only on the gross magnitude of the krill harvest, but also on where and what sizes of krill are caught, and how closely these correspond to the location and sizes of krill eaten by whales. It may also depend on the detailed population dynamics, and other aspects of the biology (feeding behaviour, etc.) of both krill and whales. This sort of research cannot be done overnight, and effective management of the Southern Oceans as a complete ecosystem will have to be based on good, long-term multidisciplinary research. CCAMLR has available to it a good scientific basis for starting its work as a result of the research that has already been carried out by the Scientific Committee on Antarctic Research (SCAR), and especially its Biological Investigations of Marine Antarctic Systems and Stocks (BIOMASS) programme.

The role of the Antarctic Treaty

In the narrowest sense the Antarctic Treaty has had very little direct impact on the management of marine resources, and Article VI, which states that nothing in the Treaty should prejudice the rights of any state with regard to the high seas, might seem to preclude such impact except for seals or penguins which can be harvested when they come ashore. In this connection one of the earliest actions under the agreed measures for the Convention of Antarctic Fauna and Flora was to give special protection status to fur seals and to the Ross seal.

It is also true that, while the responsibility under Article IX (I) (f) of the treaty for the 'preservation and conservation of living resources in Antarctica' makes no distinction between terrestrial and aquatic resources, the initial focus of those concerned with the treaty was on the terrestrial animals and plants and those aquatic animals that came to land, or the ice-shelves, to breed. For example, the Agreed Measures explicitly discuss mammals (but excluding whales) birds and plants, but make no mention of fish, krill or other animals.

Such a narrow interpretation would ignore the very significant contribution that the Antarctic Treaty mechanism has made to managing marine resources by negotiating CCAMLR and the Convention for the Conservation of Antarctic Seals, and in helping to determine the content of the two conventions and the way in which CCAMLR is likely to operate.

Though the conferences at which these conventions were finally signed took place outside the formal framework of the Antarctic Treaty, these final acts were the result of lengthy discussions and negotiations, much of which took place, formally or informally during Antarctic Treaty Consultative Meetings. As remarked in the previous section, these conventions are notable for the fact that they were established wholly or largely before there was large-scale harvesting of the resources concerned, and well before such harvesting began to have obviously harmful effects. Such public forethought is highly exceptional in the history of the utilisation of national resources, and owes a lot to the forethought and initiative of a few individuals in a number of the Treaty countries.

Such private forethought would, however, have been of little use if the Antarctic Treaty did not give it a framework in which it could be effective. Despite the initial lack of focus on marine resources, the Treaty did give its signatory countries a joint interest and a responsibility for all the resources south of 60° S latitude. Because of the nature of the marine resources, for which the 60° S latitude has no special significance, this joint interest tended to extend to the natural boundary at the Antarctic Convergence.

Governments do not like taking action unless it is very clear that failing to take action will cause much more trouble than doing something. If the Antarctic Treaty had not existed and provided both the framework and the political justification for initial negotiations, it is fairly certain that there would now be no sealing convention and unlikely that any negotiations about a more general 'ecosystem management' convention corresponding to CCAMLR would have got much beyond initial confrontations between those interested in conservation in the narrow sense of discouraging any exploitation, and those interested in immediate exploitation with not much concern for long-term interests.

The potential for confrontation are well illuminated by the recent history of the IWC. Its problems have been briefly mentioned earlier. They illustrate that a management body cannot work well unless there is a reasonable balance between different interests. In the IWC the balance has swung from undue dominance by short-term economic interests to dominance by conservation interests, and to a large extent the more extreme conservation views.

Unfortunately the whaling conflicts have spilt over into the wider discussions on the Antarctic marine ecosystems as a whole, since the member countries, and some of the individual scientists, are often the same. The conflicts in the IWC have reached a stage at which there is a great deal of mutual distrust of the objectives and scientific integrity of the other side. In particular, those countries which still engage in whaling feel, not wholly without justification, that some of those attending the IWC are only interested in stopping all whaling, without much consideration of how this is done, or whether it could be in accordance with the spirit and intention of the original convention.

The whaling countries are largely the same countries interested in harvesting krill and fish in the Antarctic. They certainly would not have been prepared to consider a new commission that looked like a repeat of the IWC. Probably it was only because of the existence of the Treaty mechanism, and as part of a Treaty initiative, that they were prepared to consider actively a new conservation mechanism for the marine resources.

The Treaty also provides a model of how a rational balance between interests might be achieved. An important principle of the Antarctic Treaty expressed in Article IX (2) is that while any state which is a member of the United Nations may accede to the Treaty, participation in consultations is dependent on serious interest in the Antarctic, as demonstrated by substantial scientific research activity. The same principle is expressed in the CCAMLR convention, in which a qualifying interest for membership of the Commission may be in terms of either research or harvesting activities, (Article VII(g) of the Convention). While there are some good reasons for believing that anybody with long-term responsibilities for conservation should in principle be open to as wide a membership as possible, including states whose current interest in the resources is more potential than real, there are equally good reasons for believing that the extremely wide membership of, for example, the IWC is not always conducive to finding a constructive solution. It can be very difficult to find constructive solutions to problems, when membership of an organisation, whether nations or individuals, is open to all, without the qualification of a serious interest in the objective of the organisation. While the founders of CCAMLR would very probably have found a solution to this potential difficulty, the way was made much easier for them by the precedent of the Antarctic Treaty, and the existence of the consultative powers as a 'club' of countries with a proven and responsible interest in the Antarctic.

In summary therefore, the contribution of the Antarctic Treaty system to current actions to manage the marine resources have been very significant. Without the Treaty the different interests would never have agreed on a convention. Further, the Treaty provided a model of the form of membership which should ensure a workable Commission. Conservation interests may well feel that the Commission is working slowly, and that the measures taken so far (minimum sizes for the mesh that can be used when trawling for fish, and minimum sizes of fish that can be landed) are more cosmetic than real. However, the Commission is moving towards more effective measures, and the collection of data that would enable these measures to be soundly based. This is clearly preferable to the only likely alternatives, which are an absence of any Commission, or a Commission dominated by conservation interests and which will be ignored by fishing nations.

Terrestrial activities

There is little interest in harvesting living resources on land. The exception is the possible harvesting of essentially marine animals when they come ashore to breed. This is a special case of the general question of the rational use of these resources, already discussed. Otherwise the conservation of living resources on land involves no serious conflict, and the task is one of ensuring that any activity carried out does not accidentally damage the resource.

This task the Antarctic Treaty, through its Agreed Measures for the Conservation of Antarctic Fauna and Flora, has done well. Three broad types of control are used – those applying to all human activities and additional protection for certain species (Ross seal and fur seal) and for certain areas. The choice of appropriate measures, and the general agreement to be bound by them, has been helped by the fact that the only activities in the Antarctic so far have been scientific research, and on a limited scale, tourism. In both cases, those involved can be expected to have an above-average interest and knowledge of the resources, and interest in their conservation. The general measures include the prohibition without a permit of the 'killing, wounding, capturing, or molesting of any mammal or bird' (Article VI), the need to 'take appropriate measures to minimise harmful interference' (Article VII), and the limitation, under Article IX, of the introduction of species that might upset the ecological balance. (In effect only sledge dogs and laboratory animals can be introduced.)

At the present scale of human activities, these general provisions are probably sufficient. They accept that nearly any activity must have some immediate impact on the ecosystem, but that all ecosystems have some natural resilience, so that if the impact is sufficiently small it will be only temporary. Thus when licences are given to kill animals as food, the numbers taken should be capable of being replaced in the following breeding season. There is, however, some belief that the Antarctic terrestrial ecosystem, at the limits of conditions under which life can exist, is less resilient than most others including the Antarctic marine system, in which the fur seal population at South Georgia has shown dramatic powers of recovery. There is little quantitative knowledge of the degree of resilience of the Antarctic terrestrial and inshore life – i.e. how much disturbance difference types of environment can withstand without permanent damage. Such knowledge, and the incorporation of this knowledge in quantitative regulations, (i.e. the amount of waste that can be discharged into coastal waters, or the number of visitors that can be allowed to a given site) may become more important if Antarctic activities increase. They would seem to be essential if industrial activities (e.g. oil extraction) ever became a practical possibility. In that case the Agreed Measures could still be seen as an important first step.

Controls for special areas which were initially for the creation of specially protected areas (SPA), were established prohibiting many activities, including driving vehicles. Some 17 SPA have been designated, and are given very effective protection. However, the strictness of the measures and especially the resultant difficulty in carrying out research in these areas, have probably lessened the enthusiasm for designating SPAs, and there has been some feeling that they do not cover wide enough areas to achieve all the purposes that such areas could serve. These purposes were identified at the Seventh Consultative Meeting, at which it was concluded that SPAs should include, *inter alia*, representative examples of the major Antarctic land and freshwater systems, and undisturbed areas to be used for comparison with disturbed areas. The same meeting also made provision for the creation of Sites of Special Scientific Interest (SSSI), with less rigorous restrictions, which seem to be of a temporary, rather than permanent nature.

While there have been some apparent problems in achieving a full and rigorous observance of all the Agreed Measures, most recently in the case of the proposed French airstrip in Adelie Land, the measures have, in general, been obeyed and have worked well. Even the exceptions seem to have involved no serious threat to the conservation of resources. They are important mainly in showing that no agreement will necessarily be obeyed unless there is some monitoring of compliance, and that the Antarctic Treaty is no exception to this.

The Role of SCAR

No discussion of conservation in the Antarctic would be complete without a mention of the special role of SCAR. Formally SCAR (the Scientific Committee on Antarctic Research) is an organ of the International Council of Scientific Unions (ICSU), which is comprised of national academies of science, rather than governments. However, the qualifications for national membership of SCAR, and for consultative status in the Antarctic Treaty – substantial research interest in the Antarctic – is the same, so that the list of members of the two groups are very similar. This list forms what is often considered as the Antarctic club, with the Treaty being the political, and SCAR the scientific arm. The achievements (or failings) of SCAR can be taken as reflecting in some way the merits of (or flaws in) the Treaty system.

The immediate contribution of SCAR has been in providing, or helping to provide, the scientific base for the three main substantive actions for conservation – the Agreed Measures under the Antarctic Treaty, the Sealing Convention, and the CCAMLR Convention. SCAR also carries out, until such time as the majority of the Contracting Parties may establish their own scientific advisory committee, the functions of compilation and analysis of data relating to sealing, and advising the Contracting Parties accordingly.

In the long run, the more important function of SCAR is to provide the coordinating mechanism for the long-term fundamental studies which are essential to effective scientific management of any natural systems. CCAMLR has its own scientific committee and the boundary between the two scientific groups is not clear. The CCAMLR committee may be expected to concentrate on the more immediate practical aspects, (e.g. analysis of commercial catch and effort data), while SCAR concentrates more on the less immediate issues, (e.g. the physiology of krill), which may (or may not) turn out to be of direct practical significance to solving management questions.

In the context of CCAMLR's ecosystem approach, the studies carried out by the BIOMASS programme deserve special mention. This programme was coordinated by the SCAR Group of Specialists on Living Resources of the Southern Ocean, and looked especially at the structure of the Antarctic marine ecosystem, and the central role played in this system by krill. Other SCAR groups, especially the Biology Working Group in relation to the Agreed Measures, and the Group of Specialists on Seals, in relating to the Sealing Convention, have played, and continue to play, important roles in conserving Antarctic living resources, both marine and terrestrial.

11

Regulated development and conservation of Antarctic resources

M. W. HOLDGATE

Introduction

The title of this chapter appears to imply that Antarctic resources must be developed. This is by no means a unanimous view. Many people regard preservation of the present regime of peaceful scientific cooperation on the vast continent south of latitude 60° S as of so high a priority, and prospects of worthwhile development of new resources there (especially mineral resources) as so remote, that they are willing to forego all proposals for development in order to preserve the gains that have been made under the Antarctic Treaty. Linked with this view, is a feeling that the Antarctic Treaty is a fragile instrument, liable to crack under the strain of competitive commercial pressure.

Although, like most people who have worked in the Antarctic, I would number myself among those who would prefer not to see any part of its magnificent environment disrupted by mining or by the construction of associated bases, for the purposes of this chapter I am going to assume that mineral resource development and, indeed, the continued development of other living and non-living environmental resources of the region, remain possibilities. I believe this is a realistic assumption, given that the Antarctic Treaty Consultative Parties have already agreed upon a series of measures to conserve the resources and environment of the region and to regulate activities that might damage it, and are currently engaged in discussions about the content of a regime to control mineral exploration and exploitation. I believe that even the opponents of development would agree with the central thesis of this chapter; if development occurs, it must be regulated in a way which builds upon and extends the achievements already secured under the Antarctic Treaty, and which attracts recognition from the wider world community. This thesis implies development that is compatible with the principles of the World Conservation Strategy, as it sustains the productivity of renewable resources and balances alternative uses, where they compete, in a fashion most beneficial to human welfare.

The failure of regulation in the past

The history of Antarctica provides some glaring examples of the consequences of unregulated development. One of the best-known followed Cook's discovery of South Georgia, with its teeming population of fur seal (*Arctocephalus gazella*), in 1774. Some 1.2 million animals were taken from the island before 1822, and a further million from South Shetland region in the same period (Bonner, 1968; Payne, 1977). Sealers similarly devastated the stocks on other sub-Antarctic islands including Kerguelen, and on the temperate islands south of New Zealand. Elephant seals were slaughtered in large numbers for oil during the same period. After a minor recovery in the late nineteenth century, followed by a resumption of sealing, the southern fur seal stocks were so depleted that hardly any individuals at all were seen until the late 1920s, when a small population was found in the Willis Islands near South Georgia. The subsequent recovery of the population on that island, (at a rate of increase of 16.8% per annum so that some 90 thousand pups were being born a year in the mid-1970s), is one of the more remarkable wildlife recovery stories of this century. It is paralleled by less spectacular recoveries of other southern stocks, and also by a considerable increase in the numbers of elephant seal which had not been so severely depleted.

The catastrophic impact of sealing occurred because the resources were 'open access'; that is they were unregulated by governments and each sealing captain found himself in competition with all the others. Although some of the more enlightened ones, like James Weddell, recognised that they were destroying a resource which with more prudence might have yielded a steady return indefinitely, it was in nobody's interests to exercise restraint.

A similar but more interesting and complicated pattern followed in the exploitation of the Southern Ocean whale stocks. This industry began in the early years of the twentieth century, and was initially shore-based with factories on South Georgia, the South Orkney Islands, the South Shetland Islands, and in the New Zealand region. Although the first station was only built in 1904, within 3 years Antarctic whaling produced more oil than the rest of the world's whaling areas put together. The British government attempted to prevent over-fishing by issuing licences for the factories in the South Atlantic region, which were all within British-administered territory. Revenues from the whaling industry were used to fund the *Discovery* Investigations which had two ships almost continuously at sea in the southern summers between 1925 and 1939. These voyages laid the foundations of our understanding of the oceanography and marine ecology of the Southern Ocean, and provided a commendable example of an attempt to lay a foundation of scientific knowledge as the basis for the regulation of an industry. Unfortunately, the development of more profitable pelagic factory ships and catcher fleets which were not tied to a limited range from a shore station allowed the focus of whaling to shift to the high seas, beyond any national controls, and this in turn led to the competitive depletion of stocks, first of Blue whales (*Balaenoptera musculus*) and then of Fin and Sei whales, in a pattern that is now well known. Recent estimates suggest that the present stocks of Blue whales are now no more than 5% of the original, with Humpback whales at 3%, fin at 21% and Sei at 54% (Laws, 1977).

As Gulland (this volume) has pointed out, the declines or collapses of seals, whales and, in the last decade, certain fin-fish stocks, in the circum-Antarctic regions are not peculiar to this part of the world. As Gulland puts it:

> virtually every commercially attractive exploited fish stock has been allowed to decline below its most productive level. These declines have not often turned into catastrophic collapses largely because of two factors – the high reproductive capacity of most fish makes them less vulnerable to sustained over exploitation than the mammals, and most fishing fleets can move to alternative resources when stocks and catch rates in one area begin to decline. The events in the Southern Ocean should therefore not be ascribed to some unusual degree of greed or short-sightedness on the part of the harvesters, but are the predictable results of unmanaged exploitation of a common-property, open-access resource.

Nonetheless, the situation was more acute in the Antarctic regions than in the Arctic, and most of the other fishery zones of the world, because in the far south there was not the framework of unquestioned national sovereignty over the land areas, nor a regional fisheries convention. Until recently, the only regulatory body was the International Whaling Commission (IWC) which, from 1946 onwards, fought a protracted rearguard action to bring the whaling industry into balance with the resource on which its future depended.

The IWC's experience with whaling did point to certain lessons (Gulland, this volume). The first was the necessity for good quantitative scientific information on the stocks of the exploited resource, so that the consequences of alternative policies could be evaluated. The IWC did not reach the stage of reasonably thorough statistical analysis until 1960. By this stage the Blue whales had virtually ceased to be exploitable, the Fin whales were on the point of catastrophic decline. The industry was now faced with a choice between immediate and drastic catch reductions, which would have meant the writing off of much investment, or the continuance of whaling for a limited number of years in order to recoup on those investments, with the virtual certainty of termination thereafter. The IWC was powerless to dictate a solution, and the industry opted for the latter course, bringing down its quotas, but not sufficiently rapidly to create optimum conditions for the recovery of the whale stocks and the perpetuation of its own future.

Regulation under the Antarctic Treaty system

When the whaling industry was going into decline, despite the IWC's efforts, a new framework for the regulation of international activities in the Antarctic region was coming into being under the Antarctic Treaty, signed in 1959. As Heap & Holdgate (1985) have pointed out, two quite distinct approaches to environmental questions are evident in the Treaty and its associated instruments, which, with the Treaty itself, are often collectively termed the Antarctic Treaty system. The first approach defines certain broad principles intended to govern the protection of the Antarctic environment from the damaging impact of present or future activities. The second identifies and guards against particular activities which could damage the environment. The first approach is thus general and non-specific; the second is specific to particular activities. Both, however, are precautionary in their approach, for measures have been promulgated in advance of serious impact from the activities they are designed to regulate.

The general approach is especially defined in Recommendation VIII-13 and IX-5 of the Antarctic Treaty Consultative Meetings. The first of these recommends that, in considering measures for the wise use and protection of the Antarctic environment, governments 'shall act in accordance with their responsibility for ensuring that such measures are consistent with the interests of all mankind'. It provides that:

> (b) No act or activity having an inherent tendency to modify the environment over a wide area within the Antarctic Treaty Area should be undertaken unless appropriate steps have been taken to foresee the probable modifications and to exercise appropriate controls with respect to the harmful environmental effects such uses of the Antarctic Treaty Area may have.

The recommendation urges cooperation with the international scientific research programme under the Scientific Committee on Antarctic Research (SCAR) to monitor changes in the environment.

The second Recommendation (IX-5) is even more specific. In it the governments recall

> their obligation to exert appropriate efforts, consistent with the Charter of the United Nations, to the end that no one engages in any activity in Antarctica contrary to the principles or purposes of the Antarctic Treaty;

AND DECLARE AS FOLLOWS;

1. The consultative parties recognise their prime responsibility for the protection of the Antarctic environment from all forms of harmful human interference;
2. They will ensure in planning future activities that the question of environmental effects and of the possible impact of such activities on the relevant eco-systems are duly considered;
3. They will refrain from activities having an inherent tendency to modify the Antarctic environment unless appropriate steps have been taken to foresee the probable modifications and to exercise appropriate controls with respect to harmful environmental effects;
4. They will continue to monitor the Antarctic environment and to exercise their responsibility for informing the world community of any significant changes in the Antarctic Treaty Area caused by man's activity.

Regulatory measures so far agreed upon under the Treaty system have been concerned with eight areas of possible human impact. In addition, there are current discussions on the regulation of mineral exploration and exploitation. These measures fall into three groups; the first concerns the protection of flora, fauna, habitats and historic sites; the second regulates the impact of human activities not directly involving the commercial exploitation of natural resources; and the third controls potential or actual commercial activities. The measures are:

A (i) The Agreed measures for the Conservation of Antarctic Flora and Fauna;

(ii) Arrangements for designating Sites of Special Scientific Interest;

(iii) Recommendations for the protection of historic sites.

B (i) Recommendation on Environmental Impact Assessment;
  (ii) Code of Conduct on Waste Disposal;
  (iii) Measures to regulate Antarctic tourism.

C (i) Convention for the Conservation of Antarctic Seals;
  (ii) Convention on the Conservation of Antarctic Marine Living Resources;
  (iii) Prospective regime for regulating mineral exploration and exploitation.

These measures in a sense grew from the initial use of the Antarctic as a resource for science. This had been pioneered internationally in the International Geophysical Year of 1957–8, which brought with it a dramatic increase in the number of personnel wintering in the Antarctic and demonstrated the immense value of international coordination and interchange of data in studying the features of this vast region. The first proposals for the Agreed Measures for the Conservation of Antarctic Fauna and Flora, concluded in 1964 at the third Antarctic Treaty Consultative Meeting, were prepared by the scientists of SCAR. They prohibited, without a permit, the killing, capturing or molesting of any mammal or bird native to Antarctica by the citizens of any of the Treaty Contracting Parties. The Measures also established procedures for designating Specially Protected Areas (SPA) and rules controlling access to them. Finally, they provided for the designation of Specially Protected Species and for the publication of statistics of animals killed or captured under permit. This was a period during which small numbers of seals were still being killed for food for men and dogs, and there was a fear that the mounting number of scientists – and other visitors to the region – could have adverse effects on the ecological balance of particular areas. Measures were designed to monitor and regulate such impacts, and to steer visitors and investigators away from samples chosen to represent the range of habitats in the Antarctic. SCAR has normally proposed the sites for designation as SPAs and also the species (now numbering 2) to receive special protection. SCAR is also the scientific advisory body recommending the establishment of Sites of Special Scientific Interest (SSSIs) under later recommendations, the latter sites being designed to safeguard areas for long-term scientific research against inadvertent interference. Taken together with recommendations for the protection of historic sites, these three groups of measures thus provide a tolerably comprehensive species, habitat and historic monument conservation system analogous with that to be found in many countries outside the Antarctic. Responsibility for the enforcement of such measures rests with the individual national expeditions, ships and stations, which have jurisdiction over their own nationals and those operating in the Antarctic under their auspices.

The recommendations on environmental impact assessment look towards the adoption of arrangements to ensure that any research activity or supporting logistic activity that could significantly affect the Antarctic environment should be subject to prior evaluation and adjustment to minimise such impacts. Not all Antarctic bases are lovely places. There have been incidences of significant local damage in the construction of logistic facilities. A requirement to conduct and publish an Environmental Impacts Assessment (EIA) would, at least, provide an incentive to reduce these to a minimum, to the credit of the nation concerned. Such a provision and the code of conduct on waste disposal, prepared by SCAR and subsequently applied to all expeditions, address the problem of pollution, contamination and damage in the ordinary conduct of Antarctic activities, and it is conceivable that these measures may be extended by other such measures to minimise risks of long-term chemical contamination and pollution from shipping moving to and from the Antarctic.

It may be argued that the Antarctic regions are so vast, and the human footholds in the shape of scientific stations and logistic support facilities are so tiny, that there is really no need for detailed measures of this kind. But the Antarctic environment is not uniformly robust in the face of human disturbance. Although some 98% of the land area is covered by permanent ice and snow on which human marks are speedily erased by the wind or by fresh snowfall, the 2% of ice-free land is very vulnerable to human interference. This is especially so in areas around the coast where there are relatively diverse complexes of soil, simple vegetation, small lakes, and large colonies of sea birds and seals, and other ice-free areas in the interior, (especially in Victoria Land which support unique, near-sterile, soils). In the Antarctic Peninsula region human footprints on moss carpets may persist for years if not decades, while the soils of the Victoria Land dry valley, once contaminated with alien bacteria, may never return to their initial state. Moreover, man and the flora and fauna tend to be in competition in the ice-free coastal areas because these are not only the richest biologically but the easiest of access by sea and the obvious places in which to construct stations and conduct research. As a small and vulnerable fraction of the total vast region they therefore merit special protection.

These areas are also the regions of greatest attraction to visiting tourists, who particularly wish to see the large colonies of birds and seals for which the area is famous. These bird colonies appear to the casual observer to be tolerant of human intrusion, but they can actually be affected in subtle but significant ways. The metabolic rate of incubating penguins, for example, can be raised by the mere presence of an observer to such an extent that their food reserves are insufficient to sustain them for their proper spell of incubation before they are relieved by their partner; eggs are therefore abandoned to predators and breeding success reduced. Disturbance of these vulnerable coastal areas, which contain much of the attractiveness of the Antarctic to tourists and much of its value to scientists, therefore tends to have a cumulative impact which is far from obvious on any short period of observation. This is one reason why measures to regulate Antarctic tourism (which research has shown tends to concentrate on the manned stations and their adjacent wildlife) have been developed. The recommendations provide for a government to determine that it will not accept visits to its stations from tour ships, and to regulate the activity of tourists when visiting stations. A reporting system has been established to monitor where tourists land elsewhere than at stations, and some documents have been compiled for the guidance of visitors.

All the preceding specific recommendations deal with activities that do not involve the destructive exploitation of the Antarctic environment or its wildlife (if the collection of occasional scientific specimens can be excluded from that description). They thus stand apart from the third group. The Convention for the Conservation of Antarctic Seals, which is a free-standing international measure although it was developed within the treaty system, had its beginnings in 1964 when the Norwegian vessel *Polarhav* undertook an exploratory sealing voyage to test the potential for harvesting crab-eater seals (*Lobodon carcinophagus*) in the South Atlantic pack ice. This species of seal is probably the most numerous in the world, with an estimated population perhaps as high as 20 to 50 million (Laws, 1977). There was thus no fear that it could be endangered by any sealing activity that was likely to develop in the foreseeable period. It was, nonetheless, considered important that a basis for regulation be established before pressure of commercial use developed. The need for caution was strengthened by the fact that, despite many years of Antarctic research, there was still insufficient knowledge, both of the size of the stock and the dynamics of the species, to allow for the management of an industry on a sufficiently precise scientific footing. Under the Convention, extremely conservative quotas of permissible harvest per annum were set, that for crab-eater seal being 175 000, for leopard seal, 12 000 and, for Weddell seal, 5000. Ross and Southern fur seals were specifically protected under the Convention (and the Agreed Measures) and closed areas and closed seasons within which all species were protected were also established.

Since the adoption of the Convention there has in fact been no commercial exploitation of Antarctic seals, but the Convention remains in place against a future time when it may be needed, and the quotas or Total Allowable Catches (TAC) in more modern parlance, can be adjusted within the Convention to take account of new scientific knowledge.

In contrast, the Convention for the Conservation of Antarctic Marine Living Resources (CCAMLR) was concluded after the development of exploitation of fish in the circum-Antarctic oceans had already depleted some stocks severely. This was also a time when there were genuine fears among marine biologists that the growth of a krill (*Euphausia superba*) fishery could threaten the basis of the Antarctic marine ecosystem, and have consequential effects on the recovery of whales as well as the stability of seal and sea bird populations. Initial fin-fish fishery and experimental catching of krill began in 1969/70, and catches of krill rose fairly rapidly to pass 0.3 million tonnes in 1978/9 and near half a million tonnes in 1979/80 (Knox, 1983). Catches have since dropped to 0.25 million tonnes (1983/4) as a result of processing and marketing difficulties.

This is far short of some of the early estimates of available krill yield, set at as high as 50–100 million tonnes per annum. More detailed research (e.g. BIOMASS 1977, 1981; Everson, 1984), indicates that it would be prudent to revise the estimates downward substantially, but we are still short of the critical scientific knowledge of stocks on which to base a judgement about the appropriate allowable harvest. This is further reason for a cautious approach.

Under the CCAMLR Convention the available scientific information is to be reviewed by scientific advisers, who will seek to define the state of fish and krill stocks and recommended catch limits which will protect stocks when they fall below the level at which they display their maximum productivity. For fin-fish, this situation has almost certainly already been reached in some areas like South Georgia, and it will be a test of CCAMLR to see whether the TACs that are established lead to practical regulation that allows fish stocks to recover. For krill the test is still far off, because current catches are obviously far short of the point at which productivity will decline below its maximum.

CCAMLR also addresses the important point of ecological consequence, by specifying principles of conservation which include the prevention of decrease in the size of any harvested population to levels below those which ensure its stable recruitment, the maintenance of the ecological relationships between the harvested, dependent and related populations of Antarctic marine species, the restoration of depleted populations, and the prevention of changes or minimisation of the risk of changes in the marine ecosystem which are not potentially reversible over two or three decades. Behind this complex language is the goal of protecting the species that depend on krill for food against consequential depletion if krill is itself harvested on a large scale.

The measures to negotiate an agreement on the regulation of mineral resource development in the Antarctic are another manifestation of the preventative approach which has characterised the working of the Antarctic Treaty system. Most experts agree that the prospects of finding minerals in exploitable quantities in Antarctica, given the immense costs of any exploratory or exploitative process among the inland mountains or in the deep, turbulent and ice-beset off-shore seas, is extremely remote (Holdgate & Tinker, 1979; Bergsager, 1983; Gjelsvik, 1983; Holdgate, 1984, Beeby, 1985). But there are, nonetheless, concerns that an unregulated scramble for any minerals that might be worth exploiting would impose intolerable strains on the cherished framework of the Treaty system, creating risks of conflict between claimant and non-claimant states, and sacrificing much that has been built up painstakingly over the years. Discussions to prevent this happening, and to ensure that negotiations are conducted in the tranquil atmosphere of the present non-commercial situation, were therefore begun well ahead of the threat (Beeby, 1985). After tentative beginnings at the Seventh and Eighth Consultative Meetings (Rec. VII-6, VIII-14), the process started in earnest at the Ninth Consultative Meeting in 1977, with the adoption of Recommendation IX-1. This Recommendation endorsed a number of principles about any regime that might be developed to regulate mineral activities. Subsequently Recommendation XI-1 (1981) laid down some broad principles of agreement, of which the following relate particularly to the environment:

1. The protection of the unique Antarctic environment and of its dependent eco-systems should be a basic consideration;
2. The regime should include means for assessing the possible impact of mineral resource activities on the Antarctic environment and determining whether such activities will be acceptable;
3. The area of the regime should encompass the continent of Antarctica and its adjacent offshore areas but without encroachment on the deep sea bed;
4. The regime should cover all mineral resource activities at every stage;
5. The regime should be open in the sense that it should include provisions for adherence by States other than the Consultative

Parties on the understanding that adhering States would be bound by the basic provisions of the Antarctic Treaty;

6. The regime should include provisions for co-operative arrangements with other relevant international organisations;

7. The regime should promote the conduct of research necessary to make the environmental and resource management decisions which will be required.

Subsequent negotiations have blocked out many of the possible features of a regime. These negotiations are still proceeding, but it seems likely that there will be agreement on a legally binding international agreement which will be linked to the Treaty but separate from it as CCAMLR is (Beeby, 1985). Like CCAMLR it will be open for accession by all interested states. Present trends in the negotiations indicate that:

i) the regulatory framework will include a central body or commission with authority relating to the whole area covered by the regime and with regulatory powers necessary to fill in the basic framework of the agreement as any mineral activities proceeded;

ii) there will be a subordinate advisory committee responsible for scientific, technical and environmental advice;

iii) there will also be a series of smaller 'regulatory committees' with responsibility for aspects of the regulations of mining in areas identified by the commission;

iv) in environmental terms the regime will be strongly conditional in that regulatory authorities would not be able to sanction mining unless environmental evaluations pointed unequivocally to a very high probability that the proposed activities would not be damaging to the environment to any significant degree;

v) monitoring, and provisions for the withdrawal of permission if unforeseen hazards arose, are likely also to be built into the regime.

The basic proposition would thus be different from that under CCAMLR; for under the latter it is assumed that fishing is permitted until and insofar as it is not restrained, whereas under the mineral resource regime activities would be prohibited unless specifically authorised.

The scope and effectiveness of the environmental regulations

This series of general principles and specific measures needs to be examined from two standpoints: the adequacy of their overall coverage and the effectiveness with which they are implemented.

It is clear that the general principles and specific measures have evolved within the Antarctic Treaty system in a fashion that meets many of the provisos of the World SCAR Conservation Strategy (WCS) (IUCN/WWF/UNEP, 1980) with its basic principle that the productive base of the earth's biosphere must be conserved even though it is legitimate to use renewable and non-renewable resources for human benefit. The strongly precautionary approach characteristic of the Treaty is entirely compatible with the WCS. It is right that possible impacts be examined before they have developed. It is desirable that the balance of any doubt is given to the resource rather than to the exploiter (in this respect CCAMLR appears less firmly grounded than the prospective minerals regime).

This does not mean that there are not gaps in the system or needs for its further development. As Heap & Holdgate (1985) have pointed out:
- the coverage of Specially Protected Areas is not fully representative of diversity of habitats of ecosystems in Antarctica;
- there are no large, designated, protected areas comparable with the large National Parks in other areas;
- there is no overall conservation strategy for Antarctica (although the aforementioned authors have recommended that one be developed and the International Union for the Conservation of Nature is interested in undertaking this work in cooperation with the Antarctic Treaty Consultative Parties);
- there are no measures dealing with pollution prevention either from land-based sources or ships in the Antarctic region, apart from the limited measures on waste disposal;
- there is little guidance within the Treaty system on the resolution of problems of competing use where, for example, scientific research, the conservation of wildlife, fishery potential, logistic facilities or shore based mineral activities might come into contact and competition.

There remains the question of effectiveness of implementation and this is crucial. For, however difficult the negotiation of international agreements may be, they achieve nothing so long as they remain on paper. How does the Antarctic record stand up?

The Agreed Measures, and related arrangements for SSSI and protecting historic sites have been enforced by national expeditions, and SCAR has proved its worth as a scientific review body. Although there have been some failures, this group of provisions can be defended as broadly effective.

The measures regarding EIA, waste disposal and tourism appear somewhat more patchy in implementation as well as coverage. Governments do appear able to protect the immediate environs of their stations from disturbance by tourists by refusing to receive visits, but there is less certainty of control elsewhere. Only a few EIAs have been done, and the results have not been widely published. This is an area that merits attention, because the openness of the process is the best guarantee that it will be done well. There do not appear to be many data on how the waste disposal recommendations are working out.

Nature, in the shape of the density and hazards of the pack ice, has so far protected the crab-eater seals that are the most likely target of the Antarctic sealer (so long as the fur seals remain protected), and consumer resistance to seal products in Europe and North America has added a new disincentive since the Seals Convention was signed: it thus remains untested. CCAMLR, in contrast, is already facing its first test – whether it can secure action to protect the over-exploited fin-fish stocks around South Georgia. In principle, CCAMLR is one of the most enlightened international measures of its kind, with its stress on the need to manage the interlocking components of the exploited ecosystem and on the need for sound scientific data as a basis for management. What is now needed is the willing implementation of its terms by all the nations involved, which in 1980 numbered eight actually engaged in the fishery and a further seventeen with varying degrees of interest (Mitchell & Tinker, 1980).

Many contributors to international discussions about the Antarctic – including representatives of non-governmental organisations – concentrate on the status of the current international regime and how it might be modified. I suggest that, from an environmental standpoint, this is not the highest priority. The acid test is the extension and implementation of the series of measures outlined above. What criteria might be looked for? There are perhaps six ingredients in an effective environmental management provision:

- it should rest upon adequate, up-to-date, scientific information about the state of whatever environmental feature or resource is under consideration;
- there should be a competent international system within which the data are evaluated and actions decided;
- the data should be published so that they can be subjected to independent assessment;
- the evaluation and implementation system should be transparent, so that the actions are seen to follow logically;
- responsibility for action should fall unambiguously on people, organisations or other agents acting equally clearly under the authority of a government or international body to whom authority has been passed by governments.
- there should be provision for monitoring and adjustment of

controls in the light of the inevitable divergence between real and predicted environmental results.

Conclusions

The substantial series of environmental measures adopted under the Antarctic Treaty must give grounds for optimism that the process can be continued and consolidated. But this needs to be done in a fashion that carries the confidence of the wider world community. The initial image of the Treaty system as a 'closed club' is changing. Increasingly, contracting parties to the Treaty who have not achieved full Consultative Party status have been able to play an active part as observers in meetings, while the CCAMLR convention is open to non-Treaty Parties, and the minerals regime convention will almost certainly be similarly structured. There is also increasing dialogue betwen the Treaty powers and SCAR, as their scientific adviser, and IUCN, as the main non-governmental organisation concerned with conservation. The way appears open for the involvement of the International Union for the Conservation of Nature (IUCN) and the United Nations Environment Programme (UNEP) in discussions on the conservation strategy for the continent. IUCN submitted a constructive critique on the conservation and development of Antarctic ecosystems to the United Nations for consideration at the General Assembly's debate in 1984 (IUCN, 1984). These are all pointers to an increasingly outward looking attitude on the part of the Treaty powers, and an increasingly understanding view of the Treaty system by outside groups. Those who have witnessed the progressive evolution of the Treaty system through the years are almost universally convinced that it is a valuable framework, with a solid record of achievements to show and capable of further adaptation and evolution. There appears to be no other basis on which to build a future acceptable to the nations with a major commitment to Antarctica. The quest for a substitute regime universally acceptable to the world community is likely to be illusory and to divert effort from the actions urgently needed now. A regulatory framework based upon the Treaty system but broader and open to other participants could well, however, provide an adequate means for conserving the environment of one of the world's most beautiful and exciting regions.

References

C. Beeby, 'The Antarctic Treaty system as a source management mechanism: non-living resources' (1985). Paper delivered to Workshop on Antarctic Treaty System, Beardmore South Field Camp, January 1985.

E. Bergsager, 'Basic conditions for the exploration and exploitation of mineral resources in Antarctica; options and precedents' (1983), Orrego Vicuña, F.

(ed.), *Antarctic Resources Policy: Scientific, Legal and Political Issues.* Cambridge University Press, pp. 167–83.

BIOMASS *Biological Investigations of Marine Antarctic Systems and Stocks* (BIOMASS). I. Research Proposals (1977). Cambridge: Scott Polar Research Institute.

BIOMASS *Biological Investigations of Marine Antarctic Systems and Stocks (BIOMASS).* II. *Selected Contributions to the Woods Hole Conference on Living resources of the Southern Ocean* (1981).

W. M. Bonner, 'The fur seal of South Georgia' (1968). *British Antarctic Survey Scientific Reports,* **56**, 1–81.

I. Everson, 'Marine interactions' (1982) 2, Laws, R. M. (ed.), *Antarctic Ecology.* London: Academic Press.

T. Gjelsvik, 'The mineral resources of Antarctica: progress in their identification' (1983), Orrego Vicuña, F. (ed.), *Antarctic Resources Policy: Scientific, Legal and Political Issues.* Cambridge University Press, pp. 61–76.

J. A. Heap & M. W. Holdgate, 'The Antarctic Treaty system as an environmental mechanism – an approach to environmental issues' (1985). Paper delivered to Workshop on Antarctic Treaty System, South Beardmore Field Camp, January 1985.

M. W. Holdgate, 'The use and abuse of polar environmental resources' (1984) *Polar Record,* **23** (136) 25–48.

W. M. Holdgate & J. Tinker, *Oil and Other Minerals in the Antarctic* (1979). Cambridge: SCAR, Scott Polar Research Institute.

IUCN *Conservation and Development of Antarctic Eco-systems* (1984). Report to Secretary-General of the United Nations, Gland, Switzerland. International Union for the Conservation of Nature and Nature Resources.

G. A. Knox, 'The living resources of the Southern Ocean' (1983). Orrego Vicuña, F. (ed.), *Antarctic Resources Policy: Scientific, Legal and Political Issues.* Cambridge University Press: pp. 21–60.

R. M. Laws, 'Seals and whales of the Southern Ocean' (1977) *Philosophical Transactions of the Royal Society of London,* **279**, 81–96.

B. Mitchell & J. Tinker, *Antarctica and its Resources* (1980). London: Earthscan.

M. R. Payne, 'Growth of a fur seal population' (1977). *Philosophical Transactions of the Royal Society of London,* **279**, 69–79.

12

Recent developments in Antarctic conservation

W.N.BONNER

It is easy to forget today that conservation is a relatively new concept in western society and that in many other parts of the world it has still to make an impact. When the Antarctic Treaty was negotiated in the late 1950s conservation was not one of the pressing issues to be included in it. It was not until the Third Antarctic Treaty Consultative Meeting (ATCM) in 1964 that a set of rules directed towards environmental protection, the Agreed Measures for the Conservation of Antarctic Fauna and Flora, was drawn up.

The Agreed Measures form a rather simple practical code of conduct having four main articles:

  i) it is forbidden to kill, wound or capture any native mammal or bird;

 ii) harmful interference with the normal living conditions of native mammals and birds must be minimised and pollution of coastal waters avoided;

iii) Specially Protected Areas may be designated where unique or outstandingly interesting species or ecological systems can be preserved; and finally,

 iv) the introduction of non-indigenous species is banned.

More than 20 years later, most practising conservation managers would see the Agreed Measures as being rather naively drawn, and indeed, in retrospect, it would have been helpful if there had been a more effective input at the drafting stage from people with practical experience of environmental protection.

The Agreed Measures were extended by other recommendations made at subsequent ATCMs, notably at the Seventh, when the concept of the establishment of Sites of Special Scientific Interest (SSSI) was introduced. Sites can be proposed to protect research where there is a demonstrable risk of interference, or where sites are of exceptional scientific interest and therefore require long-term protection from harmful interference. (The fact that these are alternative criteria is important, but not always appreciated.) Such SSSIs are each provided with a management plan that controls the uses to which the site may be put.

The responsibility for observing the Agreed Measures (like other provisions of the Antarctic Treaty) lies with the various national groups active in the Antarctic, but general conservation policy is coordinated by SCAR, the Scientific Committee on Antarctic Research, a member body of the International Council of Scientific Unions (ICSU). SCAR is made up of delegates from each of the nations active in Antarctic research. It is frequently approached for advice by the Treaty organisation and has a subcommittee which regularly reviews conservation matters, including the designation of Specially Protected Areas (SPAs) and SSSIs.

The Agreed Measures, like all the provisions of the Antarctic Treaty, apply only to the land and floating ice-shelves. At sea there was still a need for adequate conservation. The depletion of the stocks of whales was the most conspicuous insult that the Antarctic ecosystem had suffered. There had been a number of international agreements designed to control whaling, the earliest dating from 1931. The current International Convention for the Regulation of Whaling was signed in 1946 and set up an International Whaling Commission (IWC) with the responsibility of regulating the industry. The IWC was not confined geographically to the Antarctic, but by far the greatest part of its activities were concentrated there.

At the time of the Agreed Measures, therefore, a legally effective (if practically toothless) regime existed to protect whales. This was not the case for seals at sea and when in 1964 a pilot Norwegian sealing expedition visited the Antarctic, the Treaty powers recognised the potential threat to the vast Antarctic seal stocks. In 1972 the Convention for the Conservation of Antarctic Seals was signed in London.

This covers all species of seals in Antarctic waters. It does not ban sealing completely, but sets very conservative catch limits for three abundant species, and bans the taking of the other three Antarctic seals. Closed seasons and sealing reserves were created and the Convention provides a means of regulating the catch of seals should sealing ever begin in the Antarctic, which at the present seems very unlikely.

Just as Norwegian sealing rang alarm bells in the 1960s so did the start of krill fishing in the 1970s. Krill was a far more crucial organism than seals or whales since it forms the hub of the Southern Ocean food web and the prospect of a radical reduction of krill stocks, with the inevitable consequences that that would have on dependent predators, was seen to be very serious.

The Antarctic Treaty organisation responded to this threat by its most ambitious piece of legislation – the Convention for the Conservation of Antarctic Marine Living Resources, or CCAMLR. This Convention, signed at Canberra in 1980, has as its aim to ensure that harvesting will not deplete populations below the levels that ensure stable recruitment; this implies that stocks of any harvested species should be maintained near those levels which ensure the greatest annual increment – the philosophy of maximum sustainable yield. The importance of maintaining the balance of ecological relationships between harvested, dependent and related populations is recognised and the restoration of depleted populations is one of the objectives of the convention.

CCAMLR is a philosophical scientist's convention. It is certainly not a convention for fisheries managers. Its objectives cannot be faulted but it is not at all clear how it can be made to operate in practice and its achievements in controlling Southern Ocean fisheries so far have been far from perfect. Perhaps what CCAMLR needs is more time – it is still a very young convention – but many western scientists have become despondent over its prospects.

CCAMLR's failing apart, the gap in the legal framework for effective conservation in the Antarctic could be filled by the successful negotiation of a minerals regime, which is currently under active discussion by the Treaty parties. If an acceptable minerals regime can be negotiated the Antarctic would have the necessary legal structure for proper conservation, though naturally this would have to evolve to take account of such things as the advance of our understanding of conservation theory and practice, and the shortcomings of CCAMLR.

Viewed in this light the Antarctic seems very favourably placed *vis-à-vis* conservation. But this is not the view held by a very vociferous and influential body of people from a variety of constituencies. As a broad generalisation it could be said that on the one hand is the Antarctic Treaty system, and on the other is nearly everyone else. From the conservation lobby at large comes a clamour to declare Antarctica a World Park, to hand over control of the area from the Treaty to the United Nations, and to recognise Antarctica as part of the 'global commons' or the 'common heritage of mankind'.

It is not easy to say why this has come about. Possibly a very important factor was ignorance. Some of the principle bodies involved in the debate had no, or very little, practical experience of the Antarctic. The scale of the Antarctic – the fact that it constitutes a tenth of the world's surface, both land and sea – was quite unappreciated and it was generally believed that human activities there were imminently threatening the whole region.

The International Union for the Conservation of Nature (IUCN) had taken up the Antarctica question in its World Conservation Strategy (WCS), published in 1980. This firmly consigned the Antarctic, together with the open ocean and the atmosphere, to the category of global commons, to be used jointly by the world community. There was a call from within IUCN for the designation of the Antarctic as a World Park, a concept first raised in 1972 at the Second World Conference on National Parks. This was rather coolly received by IUCN and at their General Assembly in Christchurch in 1982 a resolution was passed that called for a conservation strategy for Antarctica, an appropriate form of designation for the Antarctic environment as a whole, and for specific sites within it, a significant change of position.

Meanwhile the World National Parks Congress, meeting in Bali in October 1982, stressed the need for international protected area designation for the Antarctic, ignoring the fact that the Agreed Measures had already created in the Antarctic the largest by far strict nature reserve in the world. The Congress called for priority for this over other objectives.

One significant outcome of the Christchurch meeting was a proposal by Euan Young, Professor of Zoology at Auckland and an Antarctic veteran, for a joint IUCN/SCAR meeting on conservation problems in the Southern Ocean (including Antarctica), in an attempt to develop working relationships between the scientific community and the policy makers of IUCN.

It is perhaps appropriate at this stage to review the two organisations principally involved in this exchange. They are quite different in the scope of their objectives and criteria for membership. IUCN has as its role the development and coordination of the conservation of nature and natural resources. It is a global organisation with some 500 members from 114 countries. These members include state governments, government agencies, non-governmental organisations and private individuals.

In contrast, SCAR is an organisation focussed on the facilitation and coordination of scientific research in the Antarctic. For SCAR, conservation is a subsidiary, though very important, role. For membership of SCAR, a country must be actively engaged upon Antarctic research. There are 17 national members and 7 representatives from other bodies adhering to ICSU. SCAR national delegates are appointed by their countries' National Academies of Sciences.

SCAR was at first a little hesitant to react to the proposal for a joint meeting with IUCN. SCAR had always maintained a clear separation between science and politics and it was not clear that this was the case in IUCN. Furthermore, some of the statements associated with IUCN were implicitly or explicitly critical of the Antarctic Treaty system, with which SCAR tended to identify.

However, following a definite proposal from IUCN for a joint symposium in early 1985, SCAR accepted the invitation. It qualified its acceptance by stipulating that the meeting should not become involved with the formulation of legal measures but should concentrate on reviewing scientific knowledge and the current situation with a view to identifying potential future problems.

The joint IUCN/SCAR Symposium on the Scientific Requirements for Antarctic Conservation was held in Bonn in April 1985. It was jointly convened by Martin Angel (of IUCN's Commission on Ecology) on behalf of IUCN and by Nigel Bonner (chairman of SCAR's subcommittee on conservation) on behalf of SCAR. The aim of the symposium was to review present ecological knowledge of the Antarctic continent, its surrounding waters and its offshore islands, specifically in the context of conservation needs in the region. With the recent growth of human activity in the area it was felt important to arrive at a consensus approach to conservation so that before other interests became entrenched a rational framework of environmental management could be established. The intention was to assess where problems for conservation were potentially most acute, to examine the scientific basis for the development of protected areas and other conservation measures, and to discuss how best to disseminate information about the region.

Fifteen papers were presented at the symposium and, six working groups were set up to examine various aspects raised by the participants. One of the recommendations to come from the working groups was that IUCN and SCAR should set up a working party to devise a conservation strategy for Antarctica. This was readily accepted by IUCN and by SCAR and John Beddington, of IUCN's Advisory Committee on Antarctica, and Bonner were appointed co-convenors.

Probably the most valuable product of the Bonn symposium will not be the publication of the presented papers in *Environment International*, or even the Antarctic conservation strategy that the joint working party may ultimately produce. It may well be the establishment of a formal link between SCAR scientists, with their wide experience of field conditions in the Antarctic, and IUCN experts with experience of the mechanics and legal structure of conservation. SCAR has no international lawyers and likewise, IUCN has no field scientists with Antarctic experience.

There is no doubt that the strains on the Treaty system of conservation are increasing. The popular conservation action group Greenpeace has decided to make the Antarctic its next big campaign, aiming for its declaration as a World Park where all resource exploitation would be totally banned. This campaign will appeal strongly to many Greenpeace supporters in developed countries.

To achieve its aim, Greenpeace will draw attention to the many failings of the Treaty system. The clumsy handling of the French airstrip at Pointe Geologie; the failure of the recent Consultative Meeting to agree even the most modest system of environment risk assessment; the total failure of CCAMLR to provide protection for an important fish species off South Georgia – these are all cogent arguments for criticising the present system and are, in practical terms, unanswerable.

Is it impossible to convince Greenpeace that the World Park concept, even if it were to be endorsed by the Soviet Union and Japan, may have no better chance of success than the present system? Can they understand that a coerced signature on a Treaty does not guarantee its observance in the spirit? Does majority decision-making, as opposed to the consensus system of the Treaty, ensure enforcement? Consider as an example the implications of the much-vaunted whaling 'moratorium', pushed through by the conservation lobby in the IWC. It is arguable that the only practical result of this has been that countries that were once killing minke whales in the Antarctic under scientifically determined quotas imposed by the IWC are now harvesting the same stocks under quotas which they determine themselves – a very doubtful conservation advance from the whales' point of view. A parallel situation could arise with any agreement that was forced on a superpower. Does Greenpeace recognise that UN control of Antarctic affairs would introduce a large majority of less developed and undeveloped countries into the decision-making process whose main concern would be likely to be the exploitation of available resources, rather than a regard for the absolute sanctity of the Antarctic environment? These are questions that deserve careful consideration before a campaign is launched to destroy the Antarctic Treaty system.

I firmly believe that the best recipe for environmental conservation in the Antarctic is to build on the present structure, however imperfect. It will be tragic if a well-intentioned element in Greenpeace destroys what we now have, in the hope that it will be replaced by something better, but with no certainty that that objective can be secured.

Unless the Antarctic Treaty and, particularly, CCAMLR can bring about a major change within the next couple of years (which at present seems unlikely) this is exactly what I fear will occur. The Antarctic, which in comparison with any other comparable area in the World has an excellent conservation system, is today facing its severest threat – not from the exploiters, but from those who claim to be wholly dedicated to the cause of environmental conservation.

13

Environmental protection and the future of the Antarctic: new approaches and perspectives are necessary

J.N.BARNES

The big picture

Recently at the South Pole a group of scientists, diplomats, academics, international lawyers, journalists and environmentalists had the opportunity to view a graph of the increase in carbon dioxide levels during the past 20 years – over 25%, in a straight line. The data on which that graph is based underpins the so-called greenhouse effect theory.

This chapter is founded on the belief that it is important to keep Antarctica in perspective, against the backdrop of other developments in the world community, when determining policies for its future. For example, if scientists such as those responsible for a 1984 study by the US National Academy of Science are right about the implications of a continued build-up of carbon dioxide levels and the theory of the greenhouse effect, temperatures on earth will rise by 2 or 3 deg C during the next 75 years. That could cause the west Antarctic ice sheet to melt, raising the ocean levels by a significant – even radical – degree. What energy policy would it be prudent for world leaders to follow?

If those scientists who are concerned about the phenomenon of acid rain are correct that it will continue to bankrupt various components of the global ecosphere, and that its primary cause is the increased burning of fossil fuels in vehicles, power plants and smelters, the question arises as to the conservative approach which the world community should take concerning continued reliance on such energy sources.

If the world's human population reaches 10 or 12 billion, as most projections anticipate, how important will Antarctic marine life be in providing a portion of their diet? Oil drilling could jeopardize use of

Antarctic marine living resources. How do we compare that risk to the potential value of sustainable protein from the Southern Ocean?

Some observers have expressed concern that the region's demilitarized status could be jeopardized if 'strategic' minerals are located and nations begin to compete for control over them. Is this a risk? If so, how do we weigh the comparative value of the absence of arms? Others have suggested that to open Antarctica to minerals development will jeopardize international scientific cooperation, by providing incentives to keep information secret and by diverting scientific priorities from deeper understanding to the narrower requirements of exploitation. Is this a possibility? If so, how do we weigh the value of the open scientific cooperation that flourishes in the region at present?

I raise these questions because, as we all know, a key aspect of the debate now being conducted about Antarctica's future is whether to allow minerals development. Much political and diplomatic energy is being poured into negotiation of a minerals treaty, and into preventing the United Nations from having any involvement in that negotiation.

As we all know, the Antarctic Treaty bans all arms, arms tests and military activities in the region. In particular, nuclear testing, nuclear weapons and nuclear waste are prohibited. Some visionary scientists have proposed extending this arms-free zone 1 degree north on a regular basis, eventually yielding effective global arms control. As an example of the possibility of a sane approach to the global arms race, the Antarctic offers many instructive solutions for other areas of the world. Its unique status can also provide some impetus for new initiatives in the field of arms control, which ultimately is the greatest environmental danger. The concept of nuclear winter is becoming quickly understood both by the general public and policy makers.

The Antarctic region holds many secrets of the earth's past – perhaps even keys to the history of the solar system and the universe. It may also be the best monitoring zone for global pollution; a vantage point from which to observe – and reflect upon – the poisoning of the earth's natural systems.

For the past 27 years, scientists from many nations have worked in the Antarctic as though it were a *de facto* 'preserve', not worrying objectively about 'sovereignty' or 'strategic interests'. They have cooperated in research programs that regularly reveal wondrous insights and new treasures. Recently algae, lichens and bacteria have been discovered living in symbiotic harmony in the most difficult conditions known on earth, in the dry valleys of the Ross Desert. They are living intertwined among quartz crystals, having given up their normal forms, covered by a patina formed from chemical reactions with the rock and crystals, photosynthesizing beautiful green bands in rock heretofore thought lifeless. Under that layer is another, finer vein of very light green – the only lichen yet found to be indigenous to the Ross Desert. In thick ocean ice other algae and lichens live in fragile profusion, perhaps the primary food of the krill, *Euphausia superba*, which is the foundation species of the marine ecosystems.

In the truest, most profound sense, humans must come to see each other as relatives with a common responsibility for future generations of life on earth. By life, I mean all life, not just human life. Humans have been given use of the earth in trust. However one approaches this responsibility, the willingness to peer into the future and examine the intertwined implications of our present and past actions is absolutely essential to the honouring of that trust.

If we do this, it is possible to choose arenas for honest, peaceful cooperation. Antarctica is such a place. It constitutes 10% of the earth's surface and is surrounded by oceans whose structure and dynamics are barely comprehended by humankind. Those waters feed tens of millions of whales, seals, penguins and other seabirds in a way that reflects – like a 'clear mirror', one friend terms it – the pure vibrancy of nature, of God, of DNA, working out its fate and ours in a sinuously resilient dance through time.

The Antarctic furnishes an opportunity to demonstrate our maturity. It can be a rallying point on the world's agenda for taking care of our collective heritage, the natural resource base of the earth. If we can do this, future generations will applaud our wisdom. Antarctica can be the place where 'the interests of all mankind' are carefully weighed and appropriate choices made to give up certain habits.

The role of NGOs

In 1978 a number of non-governmental environmental organizations (NGOs) formed the Antarctic and Southern Ocean Coalition (ASOC), to monitor developments in the Antarctic Treaty system and in other forums, and to facilitate the participation of NGOs in policy and decisions about what activities to allow.

Today ASOC includes over 150 organizations in more than 30 nations. The coalition tries to send a team of observers to each intergovernmental negotiating session that concerns resources, although we are excluded from any formal participation. ASOC publishes a newspaper called *ECO* during those meetings, which is then distributed around the world. ASOC has formally applied for observer status under the Convention on the Conservation of Antarctic Marine Living Resources (CCAMLR) and has informally petitioned to be invited as an observer to the minerals negotiations. The International Union for the Conservation of Nature (IUCN) embraces a wide range of governments, agencies of government with a conservation mandate, non-governmental organizations and scientists. It has developed a World Conservation Strategy for assuring a sustainable future and the protection of a wide range of environmental values. In particular, it has made a number of specific recommendations about the management of Antarctica. IUCN has proposed that an Antarctic Conservation Strategy – a management plan – be carefully drawn up for the entire region. The governments have not yet responded to this.

IUCN has been participating as an observer to CCAMLR since 1980, but has yet to be invited to attend regular Antarctic Treaty meetings or the minerals negotiations. ASOC strongly supports opening these meetings to such organizations as the United Nations Environment Programme (UNEP) and IUCN, both of which have much expertise to offer.

It is interesting that conferences sponsored by NGOs often have provided the opportunity for positive developments that are difficult to achieve in the more political settings of a formal negotiation or meeting. For example, the Center for Law and Social Policy, Oceanic Society and Center for Environmental Education held a workshop in 1980 on implementation of CCAMLR. Its recommendations have been accepted by a number of governments, and remain fully relevant today. Key government representatives participated in the workshop, in their personal capacities.

The International Law Institute in Kiel sponsored a workshop in 1983 which brought together leading diplomats, academics and environment-alists. Working in their personal capacities, the workshop marked the first time that diplomats inside and outside the Antarctic Treaty held a dialogue about their respective views of the future. A second workshop was held in Kiel in October 1985.

The recent conference sponsored by the Polar Research Board and the International Institute for Environment and Development, held on the Beardmore Glacier and attended by over 60 diplomats, academics, scientists and environmentalists, provided a novel setting for dialogue and discussion. Again, all participants attended in their personal capacities rather than as representatives of government.

The recent IUCN General Assembly provided the first opportunity for governments and NGOs to debate the French airfield project (discussed below).

NGOs have served on a number of national delegations to Antarctic meetings. They include the United States, the United Kingdom, Australia, New Zealand and Denmark. This is a salutary trend that should be expanded.

What do environmental organizations want?

Most environmental, conservation and animal welfare groups suggest that the prudent course to follow in planning the future of Antarctica is to reject developments that could jeopardize the key values of the Antarctic Treaty. They consider that the 'full protection' option would best serve the interests of the international community. Such an approach would maintain the Antarctic as an international science laboratory and wildlife sanctuary without equal, and would ensure continuation of the region's unique demilitarized status.

A primary goal should be to avoid repeating the sorry mistakes made all over the rest of the planet: 'Let's pretend we are responsible adults', is the way one friend describes this approach. That means not waiting until there are serious problems before establishing institutions with a mandate to protect the overall environment.

Antarctic Environmental Protection Agency

Many organisations have proposed creation of an international Antarctic Environmental Protection Agency (AEPA) for the region. The AEPA's primary functions would be:

- to undertake investigations and conduct assessments of proposed activities, including logistical support facilities, bases and minerals development;
- to prepare environmental regulations for all proposed activities;
- to conduct inspections and monitoring operations; and
- to report on compliance with rules and regulations directly to Antarctic Treaty governments and other appropriate international bodies.

This can scarcely be said to be a radical idea, given the size and importance of the region and the experience with similar institutional arrangements in so many countries. Yet the Antarctic Treaty Consultative Parties have reacted rather negatively.

Marine living resources

ASOC, which has monitored the intergovernmental negotiations and meetings since 1978, has urged that the fisheries of the Southern Ocean be managed in a scientific way, taking into account the possible effects of fishing on endangered whales and other species. A symbol of our concern is the Blue whale, the largest animal ever to live on this planet, which has been reduced from well over 100000 to less than 3000 in a period of 65 years. ASOC has recommended that sanctuaries be established in which krill fishing is banned – to give the benefit of the doubt to the whales. CCAMLR requires this type of 'ecosystem as whole' approach, but the member governments seem barely interested in fulfilling the commitments of that convention.

Those nations that are fishing in the Southern Ocean are most interested in krill, perhaps the key component of the Antarctic marine ecosystem. Some fisheries experts say that krill harvests could double the present fish catches in the rest of the world. They are especially interested in locating the giant aggregations of krill called 'swarms'. Yet who can say he knows why krill congregate in those huge masses? Several years ago one krill swarm was estimated to consist of 10000000 tons. That compares to the world marine fish catch of about 70000000 tons.

Such swarms are attractive to humans fishing in the region, especially given the huge fuel costs to get the boats down there and the relatively high processing costs. But what is the function of these swarms in an ecological sense? How many are there? What could happen if humans systematically fish those swarms, having located them with the miracles of modern technology. What will be the impact on the Blue whale, which probably depends on the density of the swarms for its feeding efficiency, if the swarms are disrupted by large-scale fishing?

Minerals development

Environmental and conservation organizations are almost uniformly urging that a long-term moratorium on *all* minerals activities be agreed to by governments. That would provide the world community time to consider carefully the adverse implications of oil drilling and mining, and to compare the benefits of that course to the quite different benefits of maintaining the region as a science preserve and wildlife sanctuary.

Many organizations believe Antarctica should be formally protected as a World Park or Preserve, with all mining and drilling forbidden. That was the course staked out by one far-sighted government – New Zealand – in 1975, but abandoned under pressure from more powerful countries.

As mentioned above, ASOC has monitored the minerals negotiations, putting forward many constructive suggestions for improving the draft minerals treaty. The most recent two issues of *ECO* outline those suggestions. The documents being considered by governments are not available to the public, although many have been leaked.

Need for protected areas

The existing system of Specially Protected Areas (SPAs) and Sites of Special Scientific Interest (SSSI) is limited both in scope and number. There are no protection categories for 'wilderness' values or large-scale habitat. Proposals have been made to the Scientific Committee for Antarctic Research (SCAR) regarding new categories of protected areas, but they have not yet been discussed by governments. This topic has been the main focus of the joint symposium being sponsored by SCAR and IUCN in Bonn, 22–6 April 1985.

French airfield

One current project that concerns environmental and conservation organizations now is the construction of an airfield near the French base on Point Geologie, in Adelie Land.

The Antarctic Treaty system does not presently have the institutional structure or procedures for review of actions by a member government, even if compliance with the Agreed Measures for Conservation of Antarctic Fauna and Flora is at issue. The French airfield case demonstrates this clearly. The proposal for an Antarctic Environmental Protection Agency addresses the concerns raised by the airfield project.

There are numerous other airfields, new bases and other logistical facilities now being planned that also require careful study, on a cooperative international basis, before being built. NGOs will be monitoring those developments and making appropriate recommendations to the governments.

Infractions committee

As a general matter, NGOs are proposing that an Infractions Committee be established within the Antarctic Treaty system, to consider allegations of non-compliance with agreed rules. Such an entity is essential to maintain the credibility of the evolving system of rules and regulations.

The Antarctic Fund

Some ASOC members have proposed creation of an entity to help fund appropriate research, analysis and environmental protection in the region. The Antarctic Fund would be managed by non-governmental organizations – UNEP, IUCN, SCAR and other appropriate bodies. It would help support:

- cooperative, objective environmental assessments;
- an Antarctic Environmental Protection Agency;
- management of protected areas;

– preparation of an Antarctic Conservation Strategy; and
– the research needed to underpin sound management of the Antarctic ecosystem.

Where will the money for The Antarctic Fund come from? NGOs are challenging governments to reverse their priorities, at least in the Antarctic, and to dedicate to the Fund the savings from not building certain specified weapons. How much is cooperation and scientific understanding in the Antarctic worth? One plane? One MX 'Peacekeeper' missile? NGOs are very interested in how governments do see their priorities in a time of worldwide citizen revolt against the spectacle of the arms race.

A link to global arms control

Significant nuclear arms reductions consistent with the framework of the Antarctic Treaty system could help certain countries comply with Article VI of the Nuclear Non-Proliferation Treaty (NPT).

In September 1985 more than 100 signatories to the NPT gathered in Geneva to review the lack of progress by nuclear weapons states in complying with Article VI. Many non-nuclear weapons states were sufficiently upset with the superpowers' perpetuation of the nuclear arms build-up that they considered withdrawing from the NPT.

A Comprehensive Test Ban (CTB) Treaty is specifically envisioned by the NPT, and is essential to compliance with Article VI by the superpowers. A CTB could be seen as an extension of the ban contained in the Antarctic Treaty. The on-site inspection provisions of the Antarctic Treaty could provide a model for such provisions in a CTB, and in other arms control agreements.

Conclusion

How does the human family, currently broken into many thousands of small tribes, value Antarctica? In order to reverse the momentum of the arms race, destructive energy policies and the tide of ignorance about how future generations will have to cope with our collective childishness, some large steps are needed. In the field of nuclear arms a good start was made, fortuitously, in the Antarctic. Let us do nothing to disturb that status, and endeavour to build upon it.

If there is to be a sustainable future for the rapidly increasing billions of people on earth, there must be respect for what many original peoples might term 'our mother'. The ancients would fall to their knees in thanks to the rain. We blithely turn it to vinegar. In the Antarctic we have a clear choice. It is up to us, and to our leaders.

The exercise and cultivation of foresight is necessary. Significant sums of money will be required. It is popular in many government circles these days to decry the expenditure of government funds, except for arms. What does this reflect about our priorities? About our collective expectations of the future?

Why do we live on a planet armed to the teeth with nuclear weapons? I know it is mind-boggling to think precisely about how many weapons there are. The numbers are too big to be comprehended. One friend has prepared a diagram of the nuclear submarine fleet of the United States in comparison to the total firepower expended in the Second World War. A tiny circle in the centre of the diagram represents all of the firepower of the Second World War – including the Nagasaki and Hiroshima bombs. A circle around it represents the destructive power of *one* Trident submarine. A larger circle reflects the power of the US submarine fleet. Of course, there are bombers, land-based intercontinental ballistic missiles and Cruise missiles to consider, and the similar arsenals of the USSR, UK, France and China. Imagine if we dedicated the savings from *not* building one Trident submarine to international scientific research in Antarctica!

Yes, there is money in profusion for the human family to spend on arms, and for looking deeply, suspiciously into the future for 30, 40, 50 years or more, planning for the worst case. What is dedicated to planning for the future of life? Let us use the 'clear mirror' offered by Antarctica to show us the paths toward a sustainable future for the generations of life on earth that will follow in our footsteps.

Part IV
The Antarctic Treaty regime: minerals regulation

14

Introduction

It has become common practice for informed commentators upon Antarctic Treaty Party negotiations for a minerals regime to deny the economic viability and technological feasibility of exploitation of Antarctic minerals. It is also customary for them to adopt a conservative stance when detailing the extent of mineral deposits and when making geological comparisons between the Antarctic continent and shelf and other resource-rich landmasses. The need to tell so cautionary a tale has been prompted by extravagant assertions of a minerals boom in Antarctica and by fears of an international scramble to share in the benefits of exploitation. It remains, nonetheless, difficult for interest groups outside the Antarctic Treaty system, most particularly conservationists, to accept these assurances when minerals negotiations are proceeding. Do these commentators protest too much? Why is it necessary to negotiate a minerals regime if exploration and exploitation are so unlikely? Surely the negotiation of a legal structure to manage resource exploitation implies acceptance by the Antarctic Treaty system of minerals exploitation? Will the existence of a regulatory mechanism encourage mineral exploitation and create a presumption in favour of exploitation?

Each of these questions was discussed at the British Institute's Conference. These questions pose the dilemma for international lawyers, government negotiators and interest groups as to the most appropriate stage at which to plan measures in advance of needs and problems. Indeed, it has been a signal feature of CCAMLR that it was negotiated well before irreparable damage to stocks was inflicted by over-fishing. It may be perfectly reasonable to initiate negotiations to regulate minerals exploitation in an atmosphere in which exploitation is a distant possibility. At this time, governments are less likely to perceive their vital resource or economic interests as threatened or to be subject to domestic lobby pressure. In a calm atmosphere, states are freer to debate general principles and to develop rational planning measures.

Sensible though such contingency planning may be, it often has the opposite desired effect by concentrating attention unnecessarily upon the problem. While this dilemma appears to have been understood by the Consultative Parties, their decision to move ahead with negotiations can be explained as an attempt to demonstrate responsible resource management in the Antarctic region. The decision may also reflect a fear that, as the Antarctic Treaty makes no provision whatever for the regulation of non-living resources, the Consultative Parties had no mandate within which to impose restrictions. It is for this reason that the articles under consideration by the parties include provisions which link a minerals regime with obligations under the Antarctic Treaty and CCAMLR. It was also undoubtedly a matter of concern that, in a political environment in which Antarctica is claimed to be the common heritage of mankind, the United Nations General Assembly might attempt to fill the vacuum in the Antarctic Treaty with a separately negotiated international organisation to regulate minerals exploitation and which is responsible to the United Nations. Further, there may be substance to the view that some states hope to employ the United Nations to regulate the exploitation of mineral resources and to ensure an equitable sharing of the benefits of this exploitation, while ignoring conservation interests in environmental protection and in the continued operation of a moratorium. Inadequate and tentative though measures have been within the Antarctic Treaty system to protect the Antarctic environment, these measures may prove to be the most pragmatic and effective approach when contrasted with politically vulnerable efforts through international organisations with universal membership.

For the present, the moratorium under Recommendation IX(1)8 imposes the obligation of Consultative Parties to:

> Urge their nationals and other States to refrain from all exploration and exploitation of Antarctic mineral resources while making progress towards the timely adoption of an agreed regime concerning Antarctic mineral resource activities. They will thus endeavour to ensure that, pending the timely adoption of agreed solutions pertaining to exploration and exploitation of mineral resources, no activities shall be conducted to explore or exploit such resources.

Negotiations began in the late 1970s and the seventh round took place in Paris in September–October 1985. The eighth round was hosted by Australia in Hobart, 14–25 April 1986. The Antarctic Treaty Consultative

Parties are presently concerned to deal with key issues such as where power lies in the institutions established under a minerals regime and in establishing procedures by which joint venture assistance can be given to developing countries to ensure their participation in exploitation. Australia, and all Antarctic claimant states are unified in their demand that they play a major role in any minerals licensing system. It is this central problem of sovereignty which constitutes a major obstacle to the successful negotiation of a minerals regime for, unlike CCAMLR or agreements and recommendations on fauna and flora, the mining of non-renewable resources strikes at the heart of traditional notions of state sovereignty. No state can maintain a credible claim to territorial sovereignty if it delegates responsibility for licensing and regulation of minerals exploitation within its claimed territory.

The difficulties posed by claims to territorial sovereignty are compounded by the, not unreasonable, belief by all Antarctic Treaty Parties that regulation of Antarctic resources, whether living or non-living, is best achieved by States with long-term experience of the Antarctic and with interests in all facets of Antarctic management. This view was shared by most participants at the British Institute's Conference, particularly in view of the fact that the Antarctic Treaty includes representatives from most major geographical, ideological, population and economic groups in the international community. Others objected to an assertation of jurisdiction based upon past experience and interests and pointed out that international concerns are now wider than those of the Antarctic Treaty Parties, embracing questions of oil and atmospheric pollution and notions of common heritage. A solution to these different perceptions of jurisdiction will, if at all, be achieved through political compromise and cooperation; through the political will to maintain the spirit and principles of the Antarctic Treaty itself.

In discussions during the Conference it was also observed that it was not always clear whether unfavourable forecasts on commercial mineral prospects in Antarctica were based on an absence of mineral resources, an existing glut of resources located elsewhere or technical difficulties in exploitation. It was believed that, as with the exchange of information concerning fisheries, there needed to be a more ready exchange of information through some central source of data. It was recognised that if mineral negotiations were not successful within a reasonable time – 2 or 3 years were suggested by one – an interpretation of the moratorium imposed by Recommendation IX(1)8 was that it might no longer operate. Such a view might be adopted by Japan which has already undertaken some exploratory drilling in Antarctica.

15

Antarctic mineral resources: negotiations for a mineral resources regime

A. D. WATTS

Introduction

Public interest in Antarctica is probably greater now than for many years, and seems likely to grow. Much of this interest is due, directly or indirectly, to the intergovernmental negotiations in which participating states are seeking to reach agreement upon a regime to govern the possible future search for and development of Antarctic minerals – negotiations which began formally in 1982, but for which the groundwork had been laid over the previous decade. These negotiations are still continuing. Add to this the potential economic significance of the subject-matter of the negotiations, the appeal of Antarctica and its capacity to stir the imagination, and speculation about untold wealth lying below the ice, and it is not hard to see why public interest is aroused.

The Antarctic minerals negotiations must, however, be set in a slightly different, and less glamorous, perspective. The realities can, perhaps, best be appreciated by posing, and then trying to answer, some basic questions:

    i) What minerals?
    ii) Why try to negotiate a regime now?
    iii) Why does a regime have to be negotiated?
    iv) Why should the Antarctic Treaty Consultative Parties be conducting negotiations?
    v) What are the real problems?
    vi) What are the answers?

What minerals?

In the present state of knowledge, speculation about untold mineral wealth in Antarctica, of an 'El Dorado of the ice', is no more than that – speculation. That cannot be emphasised too strongly, or repeated too often. The geology of Antarctica is something best left to others, but very much in laymen's terms, and in summary, such geological evidence as there is suggests that mineral deposits could well exist in Antarctica. So too does the geological history of Antarctica. In other areas which, together with Antarctica, formed part of the old supercontinent of Gondwanaland – such as South America, South Africa, India, Australia and New Zealand – there are significant known mineral deposits. It is not unreasonable to expect that the remaining parts in Antarctica of what were originally the same or closely associated geological structures might prove to contain similar significant deposits. Indeed, scientifically significant occurrences of many minerals have been found in Antarctica.

In case this sounds too encouraging, it is important to emphasise the difference between scientifically significant occurrences and commercially significant deposits. All finds so far are in the former category: no commercially significant deposits have been found, nor are there yet any clear indications that commercially significant deposits exist.

The possible existence of mineral deposits is one thing; finding and developing any that do exist is something quite different. The extraordinary difficulties of doing so in the environmental conditions of Antarctica can be imagined. On land, only some 2% of the continent is free of a permanent covering ice sheet; this ice is moving at speeds which vary between a few centimetres a year near the centre and sometimes to 2000 metres a year at the seaward edge; the average thickness of the ice is more than two kilometres and in some places it may be as great as five kilometres; while, offshore, to the notoriously stormy sea conditions there must be added the hazards posed by sea ice and large icebergs, which, quite apart from their capacity to do damage to vessels or installations in their path on the surface, can deeply scour the seabed, posing special dangers to any seabed operations and installations.

There is thus no more than a possibility of there being commercially significant mineral deposits in Antarctica, little firm idea of where they might be (although some areas seem more promising than others), and severe technical difficulties in finding and extracting any that might exist. It is difficult to find fault with the conclusion stated by one authority on these matters in the following terms:

> All who have dealt with this matter have come to the same conclusion: no mineral deposits likely to be of economic value in the foreseeable future are known in Antarctica. This statement is not to say that Antarctica has no mineral resources, but rather, if they exist, they have no *economic* significance today or in the near term future.

Why negotiate an Antarctic minerals regime now?

If so little is known about the mineral resources of Antarctica, and the practical prospects for developing any that might be discovered are less a matter for the twentieth century than for the twenty-first, why, it may reasonably be asked, are governments embarking now on negotiations for an Antarctic minerals regime?

The answer lies in advantages which, paradoxically, flow from the very ignorance about mineral deposits which would otherwise seem to present only disadvantages for the conduct of meaningful negotiations. Should commercially significant deposits ever be found, or even should fuller geological studies bring appreciably closer the likelihood of such deposits being found, the negotiations would become immeasurably more difficult. States, and their operators, would have vested interests in relation to particular deposits, negotiating positions would be likely to harden in furtherance of those interests, and the prospect for reaching agreement on a general regime for Antarctica as a whole would recede.

Instead, the negotiating states saw it as better to try to establish a general regime before any such developments occurred. They wanted to try to reach agreement on at least the basic issues before the practical problems became acute, rather than wait until after the event, by which time it might well be too late to embark on negotiations with any hope of success. The minerals negotiations are thus a constructive example of states acting to forestall problems, instead of merely reacting to problems which have already become acute. It may be that the regime which will be the outcome of these negotiations may never be fully operational, for example if no commercially significant deposits are ever found; but by making the negotiating effort now the negotiating states hope to avoid the possibility of potentially serious problems in future.

Why does a regime have to be negotiated?

If there is merit in trying to establish a minerals regime for Antarctica at such an early stage as this, the question may then be asked why such a regime has to be negotiated? The applicable laws, regulations, terms and conditions governing mineral activities in a state or in its offshore areas are not normally negotiated intergovernmentally, so why should matters be any different in Antarctica?

The answer can be found in one word: 'sovereignty', or – perhaps more accurately – in two: 'disputed sovereignty'. The United Kingdom's right to prescribe unilaterally the rules governing minerals activities in the United Kingdom or on the United Kingdom's continental shelf is derived from the United Kingdom's undisputed sovereignty and sovereign rights in those areas. But in Antarctica no such undisputed sovereignty exists. Sovereignty is asserted and exercised by certain states, but it is not recognised by all others.

The United Kingdom is one of those states asserting sovereignty over part of Antarctica, which has been established as a colony known as the British Antarctic Territory. In our eyes we have all the necessary rights and powers to regulate minerals activities in the British Antarctic Territory and its offshore areas. But it is an inescapable fact that many other states do not accept our rights of sovereignty in that area and accordingly, if we attempted to regulate the activities there of operators belonging to such other states, serious disputes could easily arise. Conversely, those states which do not recognise rights of territorial sovereignty in Antarctica and who therefore in principle hold to the view that their operators have a right to conduct mineral activities in Antarctica without the need for permission from any other state, have to acknowledge that were their operators to attempt to enter Antarctica on that basis the territorial state concerned would be likely to object. Furthermore, it is difficult to see on what basis those non-recognising states could on their own purport to confer on an operator a good title to minerals won offshore. An added complication is that in one part of Antarctica the claims of three of the territorial states – Argentina, Chile and the United Kingdom – overlap, so that attempts by them unilaterally to regulate minerals activities in their claimed areas would provoke objections not only from states which in general do not recognise territorial claims, but also from its other two rival claimants. Just to complete the catalogue of complexity, there is one part of Antarctica which remains an unclaimed sector.

The international complications in such a situation will be evident. From a practical point of view it is perhaps of even greater importance that no commercial operator is likely to be interested in making the investment required for engaging in mineral activities in Antarctica if, in doing so, he risks becoming embroiled in international political disputes as a result of which he is likely to find that his title to any minerals won would, to say the least, be insecure.

If mineral activities in Antarctica are ever to become a reality, sufficient political stability is a precondition and a negotiated regime is the only way to establish that stability.

Why should the negotiations be conducted between Antarctic Treaty states?

International stability is, of course, necessary not only for minerals activities. Its value goes far wider than that. The differences of view over questions of territorial sovereignty have in the past been a source of quite considerable instability, and occasionally even of conflict. If Antarctica was to be an area of peace and harmony, the destabilising factors needed to be neutralised. This was the achievement of the Antarctic Treaty, which was signed in 1959 and entered into force in 1961. It established Antarctica as a demilitarised and denuclearised area, with comprehensive inspection provisions, and allowed for freedom of scientific research. All of this was achieved within a framework which was without prejudice to any of the Parties' views on questions of territorial sovereignty, and which effectively put that contentious issue on one side for the immediate purposes of the Treaty.

The 12 states which concluded the Antarctic Treaty were those which had demonstrated their active interest in Antarctica by participating in the International Geophysical Year of 1957–8 (most, of course, had been active in Antarctica for many years previously). The Antarctic Treaty established a procedure of periodic Consultative Meetings at which matters relating to Antarctica could be discussed and, if necessary, recommendations made to governments for further action. The states attending those meetings, now known as the Consultative Parties, were the original 12 states whose active interest in Antarctica led to the Antarctic Treaty, and such acceding states as might demonstrate a similar active interest by engaging in substantial scientific research in Antarctica.

Over the years, the Consultative Parties have adopted numerous recommendations demonstrating their concern for all aspects of Antarctica. The Antarctic Treaty itself, however, did not deal directly with the development of Antarctica's resources. So far as concerned possible mineral resources, there was growing awareness among the Consultative Parties of the dangers of allowing an unregulated scramble for such resources to develop – dangers which could affect not only the political stability of the continent, but also the unique Antarctic environment. The issue was first raised at the Consultative Meeting held in 1970. At subsequent Consultative Meetings the Consultative Parties, by stages, reached agreement on the need for a minerals regime, on certain principles on which the regime was to be based, and on voluntary restraint over any minerals activities while the elaboration of a regime was making 'timely progress'. At the Eleventh Consultative Meeting in 1981, the Consultative Parties adopted Recommendation XI-I calling for the convening of a Special Consultative Meeting to prepare a minerals regime, and laying down certain essential principles on which the regime was to be based.

The involvement of the Consultative Parties in the negotiations for a minerals regime is thus part of their overall concern and responsibility for

Antarctica as a whole. It needs to be seen against the background of the development, over the previous two decades, of the Antarctic Treaty system as an accepted and established part of the international scene. The minerals negotiations do not constitute an isolated and self-contained diplomatic conference, but are an integral part of the Antarctic Treaty system. As the architects of that system, and as the states with the knowledge and experience of Antarctica, who have by their active interest over the years demonstrated their concern for the continent, the Consultative Parties – which include market economy and socialist states, industrialised and developing states, territorial states and those who do not recognise territorial claims – are best placed to develop a regime which will ensure the most effective balance between the various competing interests which need to be reconciled.

What are the problems?

From the beginning, and particularly in light of the discussions leading to the adoption of Recommendation XI-I, it was apparent that certain issues would in the negotiations call for decisions which would effectively determine the shape of the eventual regime. Without attempting to be exhaustive and, in any event, excluding various political factors which inevitably arise between such a disparate group of states, it may be useful to mention by way of illustration several of the practical problems with which the negotiations have had to deal.

To start with, should there be a framework regime, or a mining code? One possibility would be to establish a very simple regime, setting out some basic principles and providing for suitable mechanisms, such as institutions, which could over the years come to draw up the detailed provisions which would be needed. The negotiations might then formally be over relatively quickly, and a regime established; however, in reality much of the necessary work would still remain to be done, although in a different forum. That kind of 'simple' regime was perhaps at one extreme of the range of options. At the other extreme would be the elaboration of a detailed mining code, so that the regime itself, when established, would prescribe all the necessary terms and conditions for the future conduct of minerals activities.

What account should be taken of the possible impact of minerals activities on the Antarctic environment? One of the major values of Antarctica is the environment, both for its natural beauty, its scientific purity, and its unique characteristics, including its flora and fauna. If mineral activities take place, *some* impact on the Antarctic environment is inevitable. It would be possible to imagine a regime in which no steps were taken to protect the environment, operators simply being allowed to conduct their activities in whatever manner they saw fit. At the other extreme, it would be possible to require operators to avoid any impact on the Antarctic environment; this would, if taken literally, be tantamount to prohibiting mineral activities. If something between those two extremes was to be envisaged, whereby activities could take place but in such a way as to do everything possible to protect the Antarctic environment, a number of questions arise as to the way in which that aim can be best realised.

How could the different perceptions of the claimant and non-claimant states be satisfied? It would be possible to imagine a regime in which all the functions normally associated with the control of mineral activities would be vested in the states claiming sovereignty where those activities were taking place; such a regime would, however, be unlikely to be acceptable to non-claimant states. Equally, it would be possible to conceive a regime under which all the usual control powers would be vested in the state to which an operator belonged, or in some kind of international body, just as would happen if the area where activities were taking place was completely beyond national jurisdiction. Such a solution, however, would be unlikely to prove acceptable to claimant states. Minerals activities do raise in a particularly acute form the question of ownership of or sovereignty over the land where the activities take place. Some way round the apparently irreconcilable points of view of the claimant and non-claimant states was necessary.

What degree of regulation of mineral activities should be imposed? Again, to consider the extremes which were possible, regulation could either be minimal, so that operators, perhaps merely on the basis of giving some kind of notification of what they were doing, would be allowed to conduct whatever activities seemed to them commercially most promising; *or* the regime could provide for fairly comprehensive regulation and control of what operators might do.

What, if any, institutional structure should be established as part of a regime? The question of the degree of regulation established in the regime is closely linked with the kind of institutional structure which might be needed, since a complicated system of regulation would be more likely to require a complicated institutional structure than would a regime which allowed operators the maximum freedom to do as they wished. The possible institutional structure also has implications for the position of the claimant states, for in their eyes there is, ideally, little need for additional institutions since they themselves possess, in their areas, all the powers needed to regulate mineral activities. On the other hand, the non-claimant states, who could not accept that the institutions forming part of the administration of claimant states were sufficient to regulate Antarctic minerals, would need to establish institutions to carry out the necessary functions.

What are the answers?

The minerals negotiations for an Antarctic minerals regime are still in progress. Since they began in June 1982 there have been nine rounds of negotiations, the most recent ending in October 1986 in Japan. The end of the negotiations is not immediately in sight. It is therefore not possible to say with any certainty what the solutions will be to the various problems which have been noted. Furthermore, the negotiations, like most intergovernmental negotiations, are conducted in confidence, and it is not possible, therefore, to go into any detail about the way they are developing, or about the positions taken by particular delegations.

However, this does not mean that the only response to be given to the question 'What are the answers?' is 'Wait and see'. A certain amount about the course of the negotiations has already been made public in one way or another. There is also a certain amount that can properly be said, even now, about some trends which have emerged in the negotiations and which seem very likely to be reflected in the regime as it will finally be concluded.

Once the negotiations started, it was necessary at a very early stage to consider how detailed the regime should be. It now seems clear that no attempt will be made to negotiate a comprehensive mining code for Antarctic mineral resources. The lack of knowledge about Antarctic mineral resources makes it, for purely practical reasons, impossible to try now to regulate the development of them in the kind of detail which will eventually be necessary. This is a matter which must be left until a much later stage, after a regime has been established, and more is known about the mineral resources of Antarctica. That does not mean, however, that the regime will merely establish a bare framework, leaving everything of substance for elaboration later. For a regime to be a worthwhile attempt to regulate any minerals activity in Antarctica it must go further than that. It will, accordingly, almost certainly establish the main elements for the future detailed regulation of Antarctic minerals activities while providing mechanisms in which these details can be further developed in the years to come. The result should be to offer, even at this early stage, a clear picture to governments and to prospective operators of the kind of system which will govern the future development of Antarctica's minerals resources.

One of the most important elements in the operational system being developed is that minerals activities will be permitted only in accordance with the regime. Put another way, minerals activities will in principle be prohibited: they will only be allowed to take place if they satisfy the requirements, both procedural and substantive, laid down in the regime.

The regime is almost certain to distinguish between different stages of minerals activities, namely, first, prospecting, then exploration, and finally development. Different requirements will be prescribed for each of those stages.

Prospecting – the identification of areas of mineral resource *potential* – is perhaps the one activity which may well take place in the reasonably near future. It is likely to be an activity on which operators will be able to engage without prior authorisation. A major consideration in moving towards that conclusion has been the close similarity, in terms of what activity may take place, between prospecting and the scientific research which is already permitted without authorisation under the Antarctic Treaty. Such research, and equivalent prospecting, is fundamentally environmentally benign, particularly if – as seems likely – prospecting is defined so as to exclude or place limits on some of the potentially more harmful activities which could otherwise be regarded as part of the process of prospecting, such as dredging, excavation or drilling. However, although not subject to prior authorisation, even prospecting will have to comply with other requirements laid down in the regime which have as their object the protection of the environment, or other valued interests in Antarctica, and there will almost certainly be procedures available whereby any prospecting which is thought not to comply with those requirements may be questioned.

Having found an area of minerals resource potential, an operator may want to move on to identify and learn more about specific deposits which he believes may exist – in short, in the parlance of the negotiations, to explore for such deposits. At this stage the operator's activities, involving perhaps exploratory drilling or excavations, may be substantial, and the impact on the environment significant. A system of 'no prior authorisation' would not be appropriate. In fact it is at this stage that it seems that the main controls built into the regime will come into play.

A prerequisite for any application by an operator to be allowed to explore for mineral deposits will probably be that the general area, including the particular locality in which the operator is interested, should have been identified by an appropriate body within the regime as one in respect of which such applications may be submitted. This decision to open an area for possible exploration will be one of the critical decisions in the operation of the regime. It will ensure that the areas in question are areas which can effectively be treated as units for resource management purposes, and will provide an important means of ensuring that exploration does not take place in areas where there might be unacceptable risks to the Antarctic environment.

Once an area has been identified in that way as being one in respect of which exploration applications may be submitted, it will be open to interested operators to do so in respect of particular blocks within the general area. For each application which is to go ahead, detailed terms and conditions to govern the operator's activities will be prescribed.

An operator who is to make the investment necessary to embark on exploration in Antarctica may be expected to do so only if he is virtually guaranteed the right to develop any mineral deposits which his exploration may disclose. It seems likely therefore that in many respects development and exploration will be dealt with together although, given the likely delay between the start of exploration and any possible move to development, some way of reviewing details of the operation before the move to the development stage occurs is likely to be included.

Procedures on these lines will call for some institutional arrangements, although they are unlikely to involve a large and costly international organisation. The regime's institutions seem likely to include a commission, which will be the plenary body, composed of all parties to the regime actively concerned with mineral activities in Antarctica, and which will be responsible for matters affecting Antarctica as a whole; one or more regional committees, or more limited membership, set up for each of the areas in respect of which exploration and development applications may be submitted, and each responsible for matters arising within its own area, in particular any operators active in that area; an advisory committee, to offer advice to the commission and the regional committee on scientific, technical and environmental matters; and, possibly, when and if the need arises, a secretariat.

Within a system of that broad kind, proper account can be taken of the need to protect the Antarctic environment – a need which has, from the outset of the negotiations, been a major consideration for all delegations. There is, as just noted, likely to be an institution one of whose principal functions will concern environmental matters. The regime will include both general statements of principle looking to the protection of the environment and more detailed provisions taking account of particular needs, such as the protection of specific areas. It will also include procedures for giving effect to those principles and provisions, and procedures whereby in the event of any breach of the requirements of the regime – including, but not limited to, the environmental requirements – effective steps may be taken to secure compliance. It is appropriate also to recall the point made earlier, that mineral activities will be in principle prohibited, unless permitted in accordance with the regime: unregulated, environmentally damaging activities will find no place in the regime.

One major problem to which the solution is not apparent in what has so far been said is that which flows from the divergent views of the claimant and non-claimant states. Partly this is because the solution to this problem is most unlikely to be simple and clear-cut. More importantly, however, it is because at the present stage the problem is still largely unresolved, and is one of the major remaining issues. It is too early to be able to say more than that a solution leaning predominantly in one direction or the other is not likely to prove acceptable, however much it may be accompanied by a clear 'without prejudice' text like Article IV of the Antarctic Treaty. The absence of prejudice will have to flow from substance, not just from words. Whatever solution finally emerges will have to involve a balance between the claimant and non-claimant interests which proves acceptable to both. The substantive elements in that balance will be found in various parts of the regime, and perhaps primarily within the institutional and procedural framework, enabling each to conclude that appropriate account has been taken of its position.

A minerals regime does not, however, involve just the conflicting interests of the claimant and non-claimant states. Other state interests are also involved, and the regime may be expected to accommodate them in one way or another. Mention may be made in particular of the interests of those parties to the negotiations which see themselves as developing countries, and secondly, those states which have not been participating in the negotiations but which nevertheless seek to be involved in the operation of an Antarctic minerals regime. In both these areas there is still a lot more work to be done in the negotiations, but even now it can at least be said with some confidence that the regime, like the Antarctic Treaty with which it will be closely connected, is likely to be open to accession by those states wishing to do so. Acceding states are likely to have something a little more than a purely passive role in the operation of the regime, although the greater degree of participation is likely to be possessed by those states which display an active practical interest in relevant Antarctic activities.

The whole regime seems certain to be concluded in a form which will be legally binding on the states which participate in it. A treaty of one kind or another is the most likely vehicle for this. The end of the negotiations for such a treaty is not immediately in sight, and even when a treaty is finally concluded it is likely to be subject to ratification before it can enter into force. It will therefore still be several years before a regime for Antarctic minerals is in place and operational.

Crystal-gazing is often unreliable, and sometimes even dangerous. For all the uncertainties associated with negotiations which are still in mid-course, when a treaty establishing an Antarctic minerals regime is eventually concluded, it will be likely to incorporate the elements which I have outlined and on which there is already developing a growing measure of consensus. If so, such a regime will have a good chance of proving a satisfactory and enduring basis for the search for and eventual development of Antarctic minerals if – and it is a dominant 'if' – any significant deposits are ever found and world economic circumstances, and progress in technology, make their development a rational endeavour.

16

Mineral resources: commercial prospects for Antarctic minerals

F.G.LARMINIE

The subject I have been asked to discuss is commercial prospects for Antarctic minerals and a succinct summary of what follows could well be 'virtually nil'.

Another contributor, Arthur Watts, has made the point, which I repeat, that minerals occur in widely scattered outcrops in Antarctica. It is important to set this fact in the context that first, more than 95% of the continent is ice and secondly, that there is a world of difference between an occurrence and a deposit. Mapping the occurrences may indicate the existence of a deposit of sufficient size to warrant detailed investigation (for example, by drilling), with a view to possible commercial exploitation.

Let us look at what minerals are known to exist in Antarctica. In summary, from what is presently known of the exposed area in Antarctica, only coal and iron in the Prince Charles Mountains, and coal in the Transantarctic Mountains *might* be mined if they were located on an inhabited continent. There are a lot of other known or suspected minerals. You have heard mention, for example, of the mineral potential of the Jurassic Dufek intrusion which is a layered deposit with a structure and composition which some geologists believe to be analogous to the South African Bushveld complex. In the absence of any detailed information, it is legitimate to work on the hypothesis that such complexes could, given the circumstances of their geological origin and physical properties, contain cobalt, chromium, nickel, uranium, copper and magnetite. Without a full-scale exploration programme we are left with a pattern of association (or assumption!) based upon the geological setting. This leaves the basic questions unanswered. Are there minerals present? In what concentrations? What is the real extent of the deposit?

As to hydrocarbons, the oil industry has a very jaundiced view about the possible existence of large deposits of hydrocarbons onshore or

176

offshore in the Antarctic. The Treaty nations' present interest in developing a minerals regime which will control the possible exploitation of minerals and hydrocarbons has no underlying commercial motive, and does not reflect any pressure from the petroleum industry to start exploration in this area. In this connection I am encouraged to hear that the debate on a minerals regime is concerned with general principles and presages an evolutionary approach to the establishment of a convention. In the absence of any commercial interest in the area, it would be unhelpful in my view to develop a detailed theoretical solution to a hypothetical problem at this juncture. That was what happened with the deep seabed mining provisions of the 1982 Law of the Sea Convention and these have proved to be the non-event of the century (to date!). If and when deep sea mining does become a reality, it is entirely possible that much of what is enshrined in the Law of the Sea Convention will be inappropriate for the technology and techniques involved.

When John Heap came to British Petroleum in 1975 to solicit our views on Antarctic minerals, we told him that Antarctica was at the bottom of our worldwide list of areas with sedimentary basins which were potential hydrocarbon prospects, and indeed, that it only made it onto the list because, so to speak, it was there! This was at a time when the price of oil was high, there was the threat of scarcity, and there were serious doubts about the size of world oil reserves and how long they might last. Since then there has been a glut of oil. Saudi Arabia, for example, which has a production capacity of about 12 million barrels a day, is currently producing only about 4 million barrels a day. This fact indicates the extent of the decline in oil demand today and, when coupled with the presently known and available reserves of both conventional and non-conventional hydrocarbons, it must be a major factor in any consideration of prospects of looking for, let alone producing, oil in Antarctica.

So, in the first place, I want to make it very clear that whatever governments and others may suspect, the driving force behind the moves to develop a minerals regime is *not* the result of pressure from the oil industry to facilitate access to parts of Antarctica which have been identified as potential sources of hydrocarbons. Second, I would advise you to ignore totally any figures, wherever published, which suggest that there are prospective oil reserves of 6, 10, 50 or 100 billion barrels of hydrocarbons in the Antarctic. Such figures are the product of unproven assumptions and facile arithmetic. The simplistic approach employed by some geologists, (and on occasion supported by international or national geological bodies), goes as follows: a sedimentary basin contains a given volume of sediments; within that volume of sediments there exist two types of sediments (*a*) a sediment which has not yet been seen in Antarctica but which can be a source rock which could generate hydrocarbons, (*b*) a reservoir rock which has not been seen or seen only in a few limited outcrops, which is assumed to constitute a certain percentage of the total volume of sediment within the basin. That reservoir rock has a 'porosity' which is the void space within the sediment in which hydrocarbons could accumulate and this, in turn, constitutes a certain percentage of the total volume of the potential reservoir. This pore space is then equated to a particular fluid volume and it is hypothesised that this is oil and there could be, say, 100 million barrels of recoverable fluid in the pore space which, at an estimated 25% recovery factor, would yield 25 million barrels of producible oil. This is the assumed reserve. In the case of figures published for Antarctica they are based on a total abstraction without any information on source rocks and imperfect knowledge of the sedimentary succession. The only direct evidence of hydrocarbons comprises puffs of gas from seabed coring during the 1974 Deep Sea Drilling Project. The gas consisted of heavy molecular weight hydrocarbons which occur widely in sedimentary basins and which the oil industry would not regard at this juncture as anything particularly out of the ordinary as such occurrences are a regular feature of the particular type of sedimentary sequences encountered during this investigation of superficial seabed deposits.

From these preliminary remarks I now want to outline briefly what we have learnt from our experience in the Arctic. That experience represents a long period of time in terms of the level of our present knowledge and understanding of the geology of the Arctic. The first geological information about the Arctic began to come through in the latter part of the last century and the first major geological surveys took place at about the turn of the century. But the first major oil exploration involving geology and geophysics with an accompanying exploration drilling programme started 41 years ago in the Alaskan Arctic in 1944, in Naval Petroleum Reserve No. 4., and was carried out by the US Geological Survey and the US Navy.

That programme, which ended in 1953, yielded a number of small gas and oil fields. In the late 1950s and early 1960s, private industry began to explore in Arctic Alaska and in 1969 the Prudhoe Bay oil field (the largest field in North America), was discovered. All of these discoveries, large and small, are onshore.

Let us now consider offshore exploration in the Arctic, the attendant risks involved and the degree to which Arctic exploration can be extrapolated to the Antarctic. A first important and practical point to note, and one which is not immediately obvious from a study of maps on a mercator projection, is that the Arctic Ocean in a geographical sense is a 'Mediterranean' sea because it is surrounded by land and virtually all the water comes in from the Atlantic, circulates, and goes back to the Atlantic, with only a relatively small transfer across the Behring Straits. This ocean is frozen and comprises both single and multi-year ice, with the latter predominant. There is a major onshore oil field in Arctic Alaska, some small gas fields and some as yet undeveloped small oil and gas fields on the Canadian Arctic islands and offshore in the Beaufort Sea. Much of the development to date in the Soviet Arctic has involved gas rather than liquid hydrocarbons and, in general, the exploitation of Arctic gas resources on land presents fewer environmental problems. As is the norm, Arctic exploitation began onshore and, following significant discoveries there, exploration is now beginning to move offshore. Both the onshore and offshore basins in the Arctic are favourably placed at the northern edge of the developed world, with ready markets in the population centres to the south, with their established energy-dependent economies. Compare this situation with that of Antarctica, which is an uninhabited continent surrounded by sea – I shall return to this point later.

Experience in the Arctic relates to multi-year sea ice, (i.e. frozen salt water), and the properties of that ice differ very substantially from the properties of ice formed by freezing fresh water. In Antarctica you have both frozen fresh water forming the main land-derived ice masses moving seawards from the ice-cap, and a peripheral zone of sea ice which forms, expands and contracts annually. Wave action due to the effects of wind on unfrozen sea and wind drag due to the effects of wind on frozen sea ice are environmental factors that have to be taken into account when operating in the Beaufort Sea. The force generated, (even if restricted by ice failure due to ruptures), may be of the order of 10 to 55 tons. Offshore, a large structure in areas of sea-driven pack ice and icebergs can be subjected to forces in the 5–500 thousand ton range.

The iceberg or the sea ice can impinge on the sea bottom and seabed scours have been reported in very considerable water depths. Any production facilities must therefore be put far enough below the seabed to ensure their integrity is maintained in the event of any seabed scouring. To date, shallow water Arctic offshore drilling is largely based on man-made gravel islands, or a combination of gravel island and steel caisson (which requires a lot less gravel). It is important to remember that, to date, offshore operations in the Arctic have been wholly exploratory and that there are no producing offshore oil or gas fields in the Arctic.

Reverting to the geographical situation of Antarctica: in contrast to the Arctic it is a continent almost entirely covered by ice surrounded by sea and, except in the Antarctic Peninsula, a very long way from other continents in the southern hemisphere. There is no indigenous population so that any mineral resource discovered in Antarctica would have to be transported somewhere far away to be consumed. It might even have to be brought ashore first to be stored and/or processed before it was eventually exported. To make this a commercial possibility in the case of oil you would have to find a super giant field in Antarctica for indigenous production to become a reality in the next 50 years.

The Ross and Weddell Seas are areas in the Antarctic where there are large sedimentary basins which might contain hydrocarbons. However, they are in deep water and are covered by moving ice to add to the other difficulties of isolation, lack of indigenous support (people, fuel and supplies), all of which compound the environmental problems affecting exploration of such basins. Whilst the technology has developed to drill through ice at sea this has only been done in the Canadian Arctic islands where the ice is fixed between islands and can be thickened up to provide a stable, stationary drilling platform. In the Antarctic the shelf ice is moving at *c.* 2 km a year which would preclude drilling except possibly where there are ice rises. In effect these are mini ice-caps where, contrary to the general tendency, the ice flow is radial from a centre which is an essentially static area from which it might be possible to drill. However, the Ross Sea sedimentary basin is as large as France and the odds are against an ice rise coinciding with the crest of a large geological structure. If oil were discovered in commercial quantities the problems of pipelines, storage and export would then have to be solved. Subsea production is one option, but nowhere in the world at the moment is there an entirely submarine production system in operation. The technology is still being developed and could eventually be adapted for use in Antarctica once proven in ice-free areas. But in Antarctica you again come up against seabed scouring, and this is only one of the many problems which would have to be solved before the integrity of the system could be secured.

From this brief and highly selective summary it will be clear that even with present Arctic experience, there is a critical lack of knowledge. All commercial exploitation must be underpinned by scientific investigation. At present there is not much more than reconnaissance-level knowledge of the geology of Antarctica and the correlation with the geophysical work to date is imprecise. Reflection seismology only gives time contours and they have to be correlated with velocity information from the sedimentary succession below the seabed in order to establish the geology. Without boreholes the geophysical interpretation is highly speculative. Before any commercial work is undertaken in Antarctica there is need for a few stratigraphic holes to get the required information (James Ross Island is one possible site on land), and these could be part of an international scientific geological programme with the results available to everybody. Only thus can we begin to get the basic information which is a prerequisite for understanding the geology of such a poorly exposed area. This is basic science which should not be subject to considerations of confidentiality at this early stage in the investigation of the geology of Antarctica. The undue emphasis at this stage on the requirements of confidentiality, I must warn you, can be almost as destructive of the spirit of the Antarctic Treaty as considerations of sovereignty.

17

Negotiation of a minerals regime

G.D.TRIGGS

In 1984 the Secretary-General reported that 'exploration for mineral deposits has barely started in Antarctica'. The reasons are clear; there is little incentive to search for economic deposits because of the hostile environment, lack of infrastructure, significant transportation problems, and high costs of exploration and mining operations and of developing the necessary technology. It is notable, for example, that the Antarctic continent is submerged thousands of feet below sea level by an ice-cap and only approximately 2% of the entire continent is exposed.

Despite these apparently insuperable difficulties, international attention has been attracted by discoveries of natural gas by the *Glomar Challenger* on the continental shelf off the Ross Ice Shelf in 1973, interests which became more acute after the oil crisis of 1970. Indeed, spectacular claims were made for a 'Middle East' in the Antarctic, including an assertion by the *Wall Street Journal* that oil reserves reported by the United States Geological Survey almost matched 'the proven reserves of the entire United States'. Geological surveys which have been undertaken suggest a more conservative estimation of resources. Two mineral accumulations have been identified which are sufficiently large to term 'deposits'; iron in the Prince Charles Mountains and coal in the Transantarctic Mountains. Occurrences of a wide range of minerals have been recorded including: copper, molybdenum, gold, silver, chromium, nickel, cobalt, tin, uranium, titanium, manganese, lead and zinc. Predictions are based in part upon the similarity between the Antarctic continent and other southern continents of comparable structure and age. The Gondwana thesis is that these continents were once united and, hence, that it is likely that deposits in, for example, the Bass Straight area south of Australia, are to be found in the continental shelf of Antarctica. In his first report the Secretary-General concluded, however, that as mineral occurrences are small and isolated and as current world resources in more accessible countries are adequate, the likelihood of Antarctic exploration and exploitation in the near future is slight.

In contrast, predictions for exploitation have been made in relation to Antarctica's offshore continental shelf resources because they are most likely to become economically viable and technically feasible in the shorter rather than longer term. Even so, the continental shelf in Antarctica is significantly deeper than shelves elsewhere. In 1983 J. C. Behrendt and P. L. Masters reported that there were no known resources of petroleum in Antarctica, however, again on the basis of the Gondwana thesis they predict that the continental shelf off West Antarctica and the Ross Sea show promise. Special technical problems arise, however, in exploiting the Antarctic continental shelf posed by drifting icebergs, severe pack ice and bottom-scouring icebergs. Experts have pointed out that as Antarctic ice breaks up each year it may be easier to negotiate this ice than the more permanent Arctic ice and, further, that technological developments for Arctic conditions could be used in Antarctica. The Secretary-General concludes, nonetheless, that Antarctic oil exploration is not imminent.

For these reasons, it may appear surprising that the Antarctic Treaty Parties are seriously negotiating a minerals regime and have done so through Special Consultative Meetings since June 1982. The apparent urgency does not lie in imminent unilateral mineral exploitation. Rather, it lies in the fear that unless the gap in the Antarctic Treaty is filled by a legal regime for minerals exploitation, the unresolved issue of sovereignty will re-emerge to threaten the future of the entire Treaty system. This is because, unlike the regulation of Antarctic marine living resources or fauna or flora, the exploitation of minerals is perhaps the strongest manifestation of exclusive territorial sovereignty. Claimant states have been able to maintain their assertions of sovereignty in Antarctica while, at the same time, agreeing to regional management through the Antarctic Treaty system of the environment and of living, renewable resources. But to accept regulation within the Treaty system of minerals exploitation and all that this will entail, including royalties and licensing, is likely to prejudice the validity of sovereignty claims. If claimant states accept a variation on the theme of Article IV as a mechanism to enable conclusion of a minerals treaty, the questions arise whether their claims retain any legal validity and whether traditional notions of territorial sovereignty have become so attenuated in Antarctica as to leave but an illusion of credibility.

The scene was set for the negotiation of a minerals regime in Antarctica by the Consultative Parties in Recommendation XI-I which establishes the following general principles:

- the Consultative Parties should continue to play an active and responsible role in dealing with Antarctic minerals resources;
- the Antarctic Treaty should be maintained in its entirety;
- protection of the unique Antarctic environment and of its dependent ecosystems should be a basic consideration;
- the Consultative Parties should not prejudice the interests of all mankind in Antarctica;
- a minerals regime should ensure that the principles embodied in Article IV are safeguarded in application to the area covered by the Antarctic Treaty;
- any minerals regime must be acceptable and be without prejudice to claimant states or those states which neither recognise such claims nor assert any such right;
- a minerals regime should apply to all resource activities on the Antarctic continent and its adjacent offshore areas without encroaching on the deep seabed;
- the regime should include a means for assessing the impact of mineral resource activities on the Antarctic environment and for determining whether these activities will be acceptable;
- the regime should cover all mineral resource activities at every relevant stage, i.e. commercial exploration, commercial development and production;
- any states adhering to the minerals regime which are not members of the Antarctic Treaty should become bound by the basic provisions of that Treaty;
- the regime should include provision for cooperative arrangements with other relevant international organisations;
- the regime should promote the conduct of research necessary to make the environmental and resource management decisions which will be required.

These general principles have been developed through Special Consultative Meetings. A document now forms the basis of negotiations. It has been drafted and revised by the chairman of the minerals negotiations, Chris Beeby, of the New Zealand Department of Foreign Affairs. However, the currently revised provisions are not yet publicly available. A set of articles upon which negotiations have been based is appended. As these articles have undoubtedly been altered by now and as they do not represent any form of agreed document, little is to be gained from detailed discussion of their ambit. Nonetheless, much can be learned of the structure and scope of the proposed regime and of the mechanisms for minerals exploitation.

Separate negotiation of a minerals regime

It is likely that a minerals regime will be embodied in an instrument to be negotiated separately from the Antarctic Treaty, as were the Conventions on the Conservation of Seals and Antarctic Marine Living Resources. While some states have favoured a regime under the Treaty 'umbrella', akin to the Agreed Measures, most consider it essential that a separate convention be negotiated. This is to avoid the ambiguous status of the Agreed Measures, which it can be argued are no more than 'mere' Recommendations, and also to counter allegations of furtive activities by the Antarctic 'club' under the Antarctic Treaty. A separate treaty would, of course, call international attention to the claimed exclusive competence of the Consultative Parties to regulate mineral exploitation in Antarctica. Any danger which may lurk in the possibility of an adverse international response to such a claim is outweighed by the legal difficulties of negotiating an agreement under the terms of the Antarctic Treaty. Certain of these difficulties were discussed at the Nansen Foundation Meeting, including the possibility that a minerals regime would conflict with the objects and purposes of the Treaty. The policy of unanimity and the complexity already encountered of amending the Treaty to deal with non-living resources, also suggest that a separate agreement would be preferable. Finally, it is likely to be easier to encourage third states to accede to a new Treaty than to bring them into the Antarctic Treaty and to accept the Recommendations made under it.

Although a minerals regime is thus likely to be negotiated separately from the Antarctic Treaty, it may mirror certain provisions of earlier conventions. It will be necessary, for example, to preserve the rights and duties of the Parties to the Antarctic Treaty and to the Sealing and Marine Living Resources Conventions. Furthermore, a variation on the theme of Article IV of the Antarctic Treaty might be anticipated to ensure that all adherents, whether or not they are parties to the Antarctic Treaty, accept the principles of that Article.

The draft proposes such a separate regime. It establishes three administrative bodies; a commission to give effect to the objectives of the Treaty and *inter alia*, to authorise the issue of exploration and development permits; a scientific, technical and environmental advisory committee to assess information on the impact of mineral resource activities on the Antarctic environment; and a regulatory committee to draw up and enforce a management scheme for further exploitation.

The draft proposes three stages of administration. The first concerns prospecting which, because of the difficulty in distinguishing scientific research from prospecting, has yet to be defined. Article XXIII provides that a 'sponsoring state' is to notify the Commission that it intends to prospect for certain minerals within a specific general area. The commission may declare that prospecting is prohibited in certain areas. At this stage, prospectors do not acquire any rights or title to resources and, importantly, in contrast with scientific information under the Antarctic Treaty, proprietary data acquired during prospecting is not subject to disclosure.

The second stage concerns exploration. The sponsoring state is to notify the commission when an operator under its sponsorship intends to initiate an exploration programme and to include various and precise coordinates of the area of interest. The advisory committee is then to recommend to the commission whether the area should be open for exploration, whereon the commission may determine whether exploration should proceed in light of environmental risks. Once this determination has been made the sponsoring state may apply for blocks. If competing or overlapping applications are made, the advisory and regulatory committees are to resolve the conflict. On application, the advisory committee is to prepare scientific, technical and environmental guidelines including provision for cancellation of a permit for failure to comply with requirements.

A separate regulatory committee is then to be convened in respect of each exploration application. Its task is to draw up a management scheme to cover both exploration and development of the block and to give effect to measures adopted by the commission and to advisory committee guidelines. The scheme must include, *inter alia*, licensing arrangements, inspection and enforcement provisions and the payment of taxes and royalties. Once the commission has approved the scheme, it may issue an exploration permit which accords the operator security of tenure in relation to the block claimed and the right to apply for a development permit within that block.

The third stage concerns development. The sponsoring state is required to lodge with the secretariat a notification of the intent to seek a development permit. The advisory committee must then consider the adequacy of the management scheme. If no modifications are required, it may refer the scheme to the commission which 'shall, without further notice' issue the development permit.

The purpose of the draft is to provide a statement of principles for a minerals regime rather than a detailed mining code. Thus the precise procedural and regulatory provisions must depend upon the establishment, in the first instance, of the three institutional bodies. It should also be noted that the proposals cover 'mineral resources' which include 'all non-living natural non-renewable resources' such as fossil fuels and metallic and non-metallic minerals. The regime is to apply to all mineral resource activities on the continent of Antarctica and to all other land areas south of 60° S latitude, including ice-shelves and the seabed and subsoil of their continental shelves and offshore areas. The proposed regime specifically excludes encroachment on the deep seabed within the area.

Decision-making

The established practice of seeking unanimity of decisions is unlikely to be appropriate for a future minerals regime. The Consultative Parties have achieved a remarkable record of consensus when making Recommendations concerning community interests such as the environment or scientific research, but the prospects for such agreement on minerals are less sanguine. Consultative Parties and other parties to a minerals regime, are unlikely to have identical economic or political interests in mineral exploration and exploitation, as was the case, for example, in the Agreed Measures. If the principle of unanimity were to be retained in a minerals regime, it would be possible for any Party to veto a particular mining proposal. The result would be that a claimant state might be denied access to minerals within its Antarctic sector. Such a result must be unacceptable to any state wishing to maintain its claim to territorial sovereignty.

The draft abandons the principle of consensus decision-making except for certain decisions by the commission and advisory committee such as budgetary matters, the establishment of subsidiary bodies or the designation of sizes of blocks for resource exploitation. All matters of substance are by a two-thirds majority of members present and voting; other matters are by a simple majority. Decisions of the regulatory committee are also by a simple majority. Thus, although the commission decides by consensus whether a general area should be open to exploration applications, the processing and evaluation of the exploration and development permits may not be blocked by a single veto.

Preservation of positions on sovereignty

A minerals regime must balance the respective positions on sovereignty of claimant and non-claimant states. Solutions based solely either on the assumption of sovereignty (where exploitation would take place under the jurisdiction of the claimant state), or on non-recognition of sovereignty (where exploitation would take place on a free-access basis with no economic return to a claimant State), will be unaceptable to one camp or the other.

As noted earlier, it is easier to reach agreement over the exploitation of living resources, without jeopardising a claim to territorial sovereignty, than it is in relation to minerals. However, the Spitzbergen precedent suggests that, to give non-claimant states a right of access to territory, upon conditions, to exploit resources found there does not necessarily diminish the sovereignty of claimant states. Australia, for example, may thus reasonably argue that to allow other states access to mine within its Antarctic Territory, does not imply a prejudice to Australian sovereignty. The crucial point is that to allow such access, without either asserting or retaining control over the processing of applications and the issue of licences, without insisting upon receiving a percentage of royalties or other benefits, may be interpreted as evidence that a claimant state has accepted a status of less than full sovereignty.

The draft meets the issue of sovereignty simply by adopting the language of Article IV of the Antarctic Treaty which preserves the respective juridical positions of the parties. Draft Article VII provides that:

> Nothing in this regime and no acts or activities taking place while this regime is in force shall:
> (a) consitute a basis for asserting, supporting or denying a claim to territorial sovereignty in the Antarctic Treaty area or create any rights of sovereignty in the Antarctic Treaty area;
> (b) be interpreted as a renunciation or diminution by any party of, or as prejudicing, any right or claim or basis of claim to territorial sovereignty in Antarctica or to exercise coastal State jurisdiction under international law within the area to which this regime applies;
> (c) be interpreted as prejudicing the position of any part as regards its recognition or non-recognition of any such right, claim or basis or claim;
> (d) affect the provision of Article IV, paragraph 2, of the Antarctic Treaty that no new claim, or enlargement of an existing claim, to territorial sovereignty in Antarctica shall be asserted while the Antarctic Treaty is in force.

Whatever the criticisms of this and similar provisions, it has been successful in enabling the parties to avoid their conflicting positions on sovereignty and to address the more pressing needs of the environment and marine living resources. For reasons which have been explained, it may be doubted whether, as a matter of law, this device alone can be successful in preserving the interests of claimants in the context of a minerals regime. It may be, however, that claimants are prepared to accept this risk in order to achieve the immediate goal of constructing a minerals regime within the Antarctic Treaty regime and thereby to pre-empt attempts by some developing states to include Antarctica within the jurisdiction of the International Seabed Authority.

The interests of claimants or prospective claimants are to be protected primarily through membership of the regulatory committee. The proposed commission is to comprise the Antarctic Treaty Consultative Parties, and acceding Parties to the minerals treaty during such time as it, or an operator it sponsors, is 'engaged in Antarctic mineral resource activities'. The proposed advisory committee includes the members of the commission and any party to the minerals treaty which is a party to the Antarctic Treaty and has conducted scientific research relating to Antarctic mineral resources. Observer status may be accorded to both these bodies, to other parties to the Antarctic Treaty and the minerals treaty and to certain international organisations. The proposed regulatory committee is to be established according to a more complex formula. It comprises the sponsoring state which has lodged an application, the parties to the treaty which are claimant states in the block to which the application applies, the two states which have asserted a 'basis of claim' in Antarctica (being the Soviet Union and the United States), the 'requesting state', and three parties designated by the states making a claim in the block subject to the application. No more than four states which assert rights of or claims to sovereignty in Antarctica may be appointed to the committee and the chairman of the commission must ensure that the committee has an equal number of claimants and non-claimants. In this way, membership of the regulatory committee responsible for drafting the crucial management scheme protects the interests of the claimants and the Soviet Union and United States on the one hand (the 'owners') and the sponsoring or requesting states (the 'miners') on the other. While decisions are by a simple majority, any majority must include a claimant state. For this reason, claimants retain what amounts to a veto vote over the appointment of a member to draft the management scheme. It has been suggested that this vote is likely to be used to ensure that the claimant state drafts the management scheme. In this way, the special interests of the claimant can be protected, as it will be seen to control those aspects of mining which most closely impinge upon sovereignty such as licensing, taxes and royalties, depletion policy and enforcement.

It is proposed that participation in resource activities is to be limited to parties to the minerals regime or those sponsored by Parties. Activities may be carried out through cooperative ventures between parties, to by operators which may be parties, a state entity of a party or a natural or juridical person with a 'substantial and genuine' link with a party. In the case of natural persons this means nationality, and in the case of juridical persons it means that it is established under the laws of the sponsoring state and has its management and resources within that State. As a party engaged in a cooperative venture, or sponsoring an operator will necessarily be 'engaged in Antarctic mineral resource activities' within the meaning of draft article X2(a), such a party will also be a member of the commission and probably also the advisory committee.

Third states

The paramount aim of a stable and predictable regime for mineral exploration and exploitation cannot be achieved if states outside the Treaty system ignore the regime, claim an open right to use Antarctic resources and authorise their nationals and vessels to mine on the continental mainland or in adjacent offshore areas. It is also necessary to guard against the phenomenon of 'free riders'. It is thus important to ensure that any interested state with the capital and technological capacity to mine in the Antarctic, should be encouraged to accede to the regime, that wide rights of accession should therefore exist and that there should be no discrimination in determining which states will have the opportunity to engage in mining.

The draft does not include any provision on accession. It is, however, likely that the treaty will be open to accession on terms similar to those of Article XXIX of the Convention on the Conservation of Antarctic Marine Living Resources, that is, whenever a state has an 'interest' in research or exploration activities in marine living resources. Any such acceding state will then become bound by Articles I, IV, V and VI of the Antarctic Treaty through the interlinking draft Article VIII. Thus, while the Antarctic Treaty is open to all states, they must accept the basic obligations of the Antarctic Treaty including the cornerstone Article IV which preserves the position of states on the question of sovereignty.

An insuperable problem lies in proposed mineral activities in Antarctica by states which do not wish to become bound by any minerals regime. Article XIV of the draft requires Parties to 'exert appropriate efforts,... to the end that no one engages in any Antarctic mineral resource activities' contrary to the regime. This has no legal effect upon third states. Unilateral resource activities, while unlikely for the future, remain a potent threat to management within the Antarctic Treaty regime and can be avoided, if at all, only through diplomatic pressure and international opinion.

Enforcement

A minerals regime should also include an effective system of enforcement, possibly with constant monitoring at each mining site, under the authority of a supervising administrative body. As such a system might be seen to prejudice the respective positions of claimant and non-claimant states, it may be necessary to embark upon another exercise of deliberate ambiguity. Claimant states would have to delegate jurisdiction to enforce standards and hence conditions to an administrative body. Such a delegation can readily be justified on the grounds that the remoteness and size of Antarctica, and its hostile climate, make it impossible for each state to maintain a separate, effective enforcement team in the area. Accordingly, effective management, pursuant to the regime, would require a cooperative force. Delegation of power to that force need not necessarily be seen as diminishing a claimant state's territorial sovereignty.

The draft provides that each party to the regime is responsible for ensuring compliance. All stations, installations and equipment relating to mineral resource activities and all ships and aircraft supporting these activities are to be subject to the same inspection obligations as applies under Article VII of the Antarctic Treaty.

On the assumption that the claimant state will be the state designated to draft the management scheme, it will prescribe the law applicable to the operator, the system of inspection and enforcement and the conditions in which an exploration permit may be suspended, modified or cancelled. While there has been some delegation of the enforcement jurisdiction, claimants will retain direct control over exploration within their claimed territory through the management scheme. It should be noted in this context that the power of the advisory committee to review and modify the management scheme, prior to authorisation of a development permit, excludes the power to make alterations to the inspection and enforcement provision in draft Article XXX(c). It also excludes any right to alter provisions in the scheme for reimbursement by the operator of the costs of implementation of the scheme in draft Article XXX(c). Both exclusions appear designed to protect the interests of claimant states, again on the assumption that they will be responsible for drafting the scheme.

Distribution of the benefits

Some provision must be made for allocation of the benefits of mineral exploitation, including the profits and the revenues from royalties and taxes and participation in ancillary activities such as transport, refining and marketing. Both claimant and non-claimant states will have to accept that under any cooperative minerals regime they will be unable to gain all those benefits which would accrue to them if their respective legal positions on sovereignty were recognised by other parties. A compromise would give some benefits at the expense of others. In the absence of a minerals regime, however, it is questionable whether any benefits could be gained at all because political, and possibly legal, constraints will otherwise persist.

It may thus be necessary, for example, to provide for a percentage of revenue to be held for distribution to developing states on an equitable basis. This may deflect, if only temporarily, mounting pressure to apply the common heritage concept to Antarctica. Although the 'interests of mankind' are not, in so many words, protected by the Convention on Marine Living Resources, they have been incorporated in Recommendations under the Antarctic Treaty relating to mineral exploration and exploitation. This suggests that they might be part of a future minerals regime. Whether this will have the effect of reassuring that certain benefits of mineral exploitation will be shared with less developed states is presently under discussion by the Antarctic Treaty Parties. Such sharing of benefits need not imply a diminution in sovereignty of claimant states. If they receive a proportion of the benefits in a way which they can argue constitutes recognition of the status of the territory in which the minerals were won, their claims to sovereignty would be unimpaired.

The draft provides in Article XIII(P) that it is the function of the Commission to:

> establish measures to ensure participation by the international community in possible benefits derived from the regime.

To provide that benefits are to be available to the international community gives credence to those who argue for the application of the concept of a common heritage in Antarctica. This provision is to be contrasted with the power of the regulatory committee under draft Article XXX(e) to establish the terms for payment of taxes and royalties in the management scheme. As a separate regulatory committee is proposed for each application which drafts a scheme suitable for the particular development permit, it is possible that taxes and royalties will vary in each scheme. It is reasonable to assume that some attempt will be made to ensure that claimant states gain special benefits, though these may be disguised in the form of administrative costs. In these ways the regime can ensure that the varied interests of the international community, claimant states and non-claimant states in the benefits of minerals exploitation can be accommodated.

Protection of the environment

It has been noted that the Antarctic Treaty makes no provision for non-living resources. It does, however, provide that the parties may agree upon measures for the preservation and conservation of living resources in Antarctica. The parties have also by recommendation, recognised their 'prime responsibilities for the protection of the Antarctic environment from all forms of human interference'. This 'prime responsibility' is assumed in the draft. The draft states the objective of the regime as being to assess the impact of mineral resource activities on the Antarctic environment and to determine whether such activities are acceptable. Various environmental 'principles' are adopted to implement the objective including the need to ensure that activities do not significantly or irreversibly change the distribution, abundance or productivity of living resources. Any environmental decision is to be based on adequate data and, when a sponsoring state notifies the commission of an intention to explore, the notification must include an environmental impact assessment along with a statement of means to mitigate any such impact. Further, the control function of the advisory committee is to provide a forum for consultation and cooperation concerning ecological, technical and other information to assess the impact of mineral resource activities on the environment.

These and other provisions in the draft demonstrate that negotiating states recognise it is critical that the Antarctic environment must be adequately protected if they are to retain management control of Antarctic resources.

Jurisdiction of the International Seabed Authority in Antarctic offshore areas

It has been observed that the exploitation of mineral resources in Antarctica is most likely to take place in the foreseeable future, if at all, on the offshore continental shelf. Indeed, resource interests in Antarctica, whether the traditional whaling or sealing or more recent krill fishing, have been essentially maritime activities. Most recommendations and inter-linked conventions relate to the marine environment including the Agreed Measures, and the conventions on Antarctic Seals and Marine Living Resources (CCAMLR). Just as the customary and treaty law of the sea have been applied by claimant states in relation to their Antarctic territories, so the 1982 Convention on the Law of the Sea has direct effect upon marine resource exploitation in Antarctica. Indeed CCAMLR and the draft minerals regime each assimilates the developing law of the sea.

Nonetheless, considerable legal and political confusions exist over jurisdictional rights, some of which have been considered in Part II.

Some states outside the Antarctic Treaty regime argue that, as there are no valid claims to territorial sovereignty in Antarctica and there are no special rights attached to the Antarctic Treaty Parties, the jurisdiction of the International Seabed Authority extends to the Antarctic continental limits, including the Antarctic deep seabed and continental shelf. It is thus the question of jurisdictional rights over non-living resources of the Antarctic continental shelf that raises the spectre of conflict between the international community as represented in the United Nations General Assembly and the Antarctic Treaty Parties. However, there is an 'Alice in Wonderland' air of unreality about such a dispute, as the 1982 Convention on the Law of the Sea may not come into force for some time and the International Seabed Authority is highly unlikely to turn its attentions to Antarctic continental shelf resources for some years, if at all. Nonetheless, the point of principle is a significant one which is raised directly by current negotiations for a minerals regime.

It is likely that the reluctance of all of the Consultative Parties to concede a role to the International Seabed Authority in the area south of the 60° S latitude will hasten the negotiation of a minerals regime in Antarctica. In relation to the oil and gas resources of the continental shelf, a minerals regime would conflict with the Law of the Sea Convention if the International Seabed Authority were to accept the view of non-claimant states that territorial sovereignty does not exist in Antarctica. However, for the short term, the Consultative Parties may fill the vacuum on the basis of their assumed responsibility for Antarctic matters, thereby purporting to demonstrate their responsible and effective management of the zone and its resources.

At the Eleventh Consultative Meeting the Parties agreed that a minerals regime is to apply to the Antarctic continent and its adjacent off-shore areas without encroachment on the jurisdiction of the International Seabed Authority in the deep seabed. This is now mirrored in the draft.

Remaining issues

A minerals regime will need to include a dispute settlement procedure. It is expected that this will be through consultation and, ultimately, compulsory arbitration. A 'fast-track' or summary procedure by a single arbitrator is presently being considered to deal with urgent risks to the environment. Amendment and withdrawal procedures are also necessary, though no review is contemplated. Further consideration must also be given to joint venture arrangements to give developing states an opportunity to participate in resource exploitation. Detailed rules and policies relating to the environment remain to be drafted and, indeed, these aspects are vital to the success of the regime.

Conclusion

As the Antarctic Treaty Consultative Parties proceed to negotiate a minerals regime, *Greenpeace* has voyaged to the Southern Oceans and declared the Antarctic continent a 'world park'. It is unfortunate that the public positions of environmental interests and the Antarctic Treaty Parties appear to be irreconcilable. Certainly negotiation of a minerals regime implies the future exploitation of non-renewable resources. In this sense, those who argue for a permanent moratorium on minerals development are necessarily vehemently opposed to States within the Antarctic Treaty system. It is nonetheless regrettable that the Antarctic Treaty Parties should appear to have lost the moral upper ground; for the history of the Antarctic Treaty regime is testimony to the dominant concern for conservation and preservation of the Antarctic environment, free access for scientific research and peaceful uses of the continent. Indeed, the interests of environmentalists, scientists and governments in Antarctica have been similar. Today, however, prospects of resource riches in Antarctica have the potential to sour and distort the achievement of the Antarctic Treaty Parties. These Parties must now demonstrate flexibility and political understanding if they are to ensure the continued rational management of Antarctica within a reformed and expanded Antarctic Treaty system.

Part V
Whither Antarctica?
Future policies

18

Introduction

An objective of the British Institute's conference, Whither Antarctica?, was to consider the future of Antarctic management. The following three papers, each of which recognises that the Antarctic Treaty Parties must make accommodation to external or outside interests, provide a stimulus for some new thinking about Antarctic regulation. With the successful negotiation of the 1982 Convention on Law of the Sea, some states have attempted to transfer the concept of common heritage of mankind to Antarctica. The assertion that Antarctica is the common heritage of mankind is, as yet, little more than a political aspiration. It has little legal content in so far as it is applied in Antarctica and it gives rise to confusion as to its consequences. Zain-Azraai moves beyond rhetoric by presenting the moral and legal bases for wider international participation in decision-making for Antarctica. He explains the dilemma of non-Treaty states which have the opportunity of joining the Antarctic Treaty System. To join is to imply that reform is possible from within the system while some non-Treaty Parties believe that radical changes can be achieved only through a wider international forum. J. A. Heap recognises that the Antarctic Treaty system is, in principle, no more or less effective than other international agreements. However, his experiences in Antarctic management have convinced him of the international 'fact of life' that a wider universal regime will represent the lowest common denominator and that regulations will be honoured in the breach rather than in observance. For this reason, he favours internal reform through which the interests of all states can be accommodated. J. R. Rowland canvasses alternative forms of regulation but returns to the immediate problem of minerals exploitation by recommending a prohibition by declaring Antarctica, not a world park, but a world conservation area. In the last resort he suggests that the time-honoured Antarctic principle of postponement might, after all, be useful.

It is significant that each of these contributors has had considerable experience in international organisations and in international diplomacy. Each takes a different approach to the future of the Antarctic Treaty system and of international participation in Antarctic affairs.

The Thirteenth Consultative Meeting in 1985 confirmed that reform within the Antarctic Treaty system is not only possible, but that it is taking place in an evolutionary way. This process might have been allowed to be continued, thereby demonstrating the flexibility and dynamism upon which all mechanisms for regulation depend for success, were it not for the possibility of minerals exploitation. Unlike living resource exploitation, mining a non-renewable resource raises moral and political issues which have thrust the entire Antarctic Treaty system into the international limelight. Despite this scrutiny, if Parties can demonstrate responsible management of the Antarctic environment and a sympathetic response to international concerns, it may be possible to convince non-party states that more radical alternatives are unnecessary.

19

Current and future problems arising from activities in the Antarctic

J.A.HEAP

Introduction

The more important current activities in the Antarctic are:
i)   scientific research (that is scientific observation and measurement);
ii) science logistics (ships, aircraft and stations and their respective activities necessary for the pursuit of scientific research in Antarctica);
iii) tourism (activities of all those whose primary purpose in being in Antarctica, whether they paid for the privilege or not, is aesthetic enjoyment); and
iv) fishing exploitation of any Antarctic marine living resources.
Among possible future activities are:
i)   exploration for Antarctic mineral resources and their possible exploitation; and
ii) the removal of ice in the form of icebergs to provide water in areas of the world where there is not enough.
One of the primary purposes of the Antarctic Treaty states is to ensure that 'Antarctica shall be used exclusively for peaceful purposes' and that it 'shall not become the scene or object of international discord'. In order to secure the first of these objectives, the Treaty provides for the demilitarisation of the Antarctic, for the provision of information about national activities in Antarctica and for international inspection of national activities to ensure that the uses to which Antarctica is being put are exclusively peaceful. These provisions of the Treaty have proved effective. Inspections have been carried out and no activities have been found which are inconsistent with the principles and purposes of the Treaty. None of the current problems in Antarctica arise out of conflict between the inherent nature of the activities going on there and the

provision for exclusively peaceful use. The Antarctic Treaty is likely to remain effective in this respect for as long as the Treaty System continues to be looked upon by the parties to it as the preferred means of managing Antarctic activity. Militarisation of Antarctica will only occur if, for some reason, the Treaty collapses. The act of militarisation of Antarctica is most unlikely, itself, to be the cause of such a collapse.

If the current and possible future activities are exclusively peaceful in their purpose, where are the problems? A definition is needed of what is meant by a 'problem'. For the purposes of this chapter I will adopt a definition which derives conveniently from the second part of the preambular paragraph of the Antarctic Treaty to which I have already referred: it is that a problem arises when an activity in Antarctica gives rise to an assertion that the Antarctic has, or might, become 'the scene or object of international discord'. 'Discord' is here used in a wide sense to include substantive disagreement.

There is an important distinction to be drawn at this point. None of the current and possible future activities described above give rise to assertions that they are not peaceful activities. Problems arise out of some aspect of the way in which the activity is conducted. No-one is saying that science, science logistics, tourism, fishing or mining are, of themselves, wrong; such problems as there are arise out of the manner in which such activities are pursued.

Moreover, if we assume that no-one is so quixotic as to pursue an activity in the Antarctic in a manner which gives rise to problems or difficulties in his own eyes, then a problem arises only when someone else asserts that he does not like the manner in which the activity in question is being pursued. The fact that almost all activity in the Antarctic is closely associated with governments means that dislike of the manner in which an activity is being pursued is usually expressed government to government and gives rise therefore to the possibility of international disagreement and discord.

Let us now look at a few problems that have arisen from current activities or that might arise from future activities, using the categories with which this paper began.

Science

As far as I am aware, no-one has challenged the scientific activity of another nation as being something which should not be done or which should be done in some other way. Some countries, notably the United States, have gone to considerable lengths to ensure that such challenges do not arise. The procedures gone through by the United States to ensure, for example, that their scientific programmes involving drilling through the Ross Ice Shelf and drilling in the Dry Valleys would not cause unacceptable environmental impacts, are cases in point. The process of observation and measurement involved in scientific research has not, of itself, and in that restricted sense, given rise to problems; nor is it likely to do so in the future as long as reasonable forethought is exercised.

Science logistics

The same, unfortunately, cannot be said about logistic support for science. The most celebrated case is that of the French plans to build an airstrip at Pointe Geologie in Terre Adelie. These plans and associated preliminary activities have led to allegations by non-governmental organisations that France is in breach of its obligations to its Treaty partners. I am not convinced that such allegations have been substantiated. That is not, however, the point: it is not so much their truth or otherwise which is of interest for the purpose of this chapter; it is rather the question of why the matter came to be seen as a 'problem'.

French plans to build an airstrip at Pointe Geologie in order to provide improved logistic support for scientific work on the continental ice sheet have been known since the late 1970s. They became known because France so informed their partners in Scientific Committee on Antarctic Research (SCAR) in the course of discussion of an international cooperative air transport arrangement then under study. There was no objection from their fellow governmental operators at that stage, nor has there been any since. Objection among non-governmental organisations erupted when it became apparent that the construction of a crushed rock airstrip would involve cut and fill to join a number of small islands, and that this work and the subsequent operation of the airstrip would have an impact on the local bird populations. Allegations by non-governmental organisations of disregard of Treaty obligations were coupled with assertions by the same organisations that the Treaty system was unable to cope with infringements of its own rules. The main thrust of their argument, however, was aimed at a more important political target. Its objective was to cast doubt on the adequacy of the Treaty system as a regulatory mechanism; if the Treaty system was unable to cope with allegations of a *prima facie* case of infringement, how could it be trusted to deal with infringements of rules that might be established for the regulation of mineral activity when national economic interest and the massive power of multinational corporations would be involved. The point is a good one which goes to the heart of the Treaty system and to the heart of regulating activity anywhere in the world on the basis of international agreement. The

Antarctic Treaty system as a means of regulating human activity is, in principle, no more nor less effective than other international agreements. In practice, however, it has the advantage of allowing for on-site inspection. Used with more vigour than hitherto this could be a means for dealing in part, at least, with problems of a similar nature.

Tourism

Tourism has not given rise to problems except when accidents have occurred. A number of Antarctic Treaty Recommendations provide adequately for Antarctic scientific stations to avoid prejudice to their activities from tourist activity. Problems have arisen when tourist ships or aircraft get into difficulties causing massive diversion of science logistics from its planned purpose to an emergency role.

Fishing

The problems arising from the exploitation of Antarctic marine living resources are those that have become familiar in connection with the exploitation of most open-access resources where short-sighted over-exploitation has led to the ruination of the resources. I would go so far as to assert on the historical evidence of the exploitation of Antarctic marine living resources that all we have learned over the last 160 years is how to prolong the agony. In 1820, exploitation of southern fur seals began; 5 years later they had been virtually wiped out as an economic resource. In 1904 exploitation of Antarctic whales began; 80 years later all but one species has been virtually wiped out as an economic resource. The only difference between the exploitation of the seals and the whales is that the whales were exploited under an agreement which required that catches should be decided in accordance with the best scientific evidence available. But this story is not unique to Antarctica, it has been repeated frequently throughout the world. If krill proves to be an economically exploitable resource, the activity will be on a grand scale and past experience gives no grounds to expect that catastrophe will not, despite the Convention on the Conservation of Antarctic Marine Living Resources (CCAMLR), be the eventual outcome – it will just take a little longer.

You will say that I am unduly pessimistic. My pessimism arises from what the record suggests about our inability to learn from past mistakes in the international regulation of open-access resource exploitation. I can view with relative equanimity the wiping out of the fur seals in the 1820s. The sealers knew no better and lived in a world where unregulated competition was the rule. What disturbs me is that now we claim to know better. We recognise the need for conservation and we call in science to our aid. But science has been subverted and this has led only to the prolongation of the agony. The subversion of science consists of a political inability to take into account the implications of the margins of error which inevitably attach to any prediction as to how nature will react to a given stimulus. Fishery scientists have consistently predicted the maximum sustainable yield of a given stock of fish in terms of a range from a higher figure, based on a series of optimistic assumptions about the population dynamics of the stock, to a lower figure based on a series of pessimistic assumptions. By so doing, the fishery scientists have been true to the limits of what they know and true to the margins of error inherent in what they know and what they can predict. The ensuing tragedy has arisen from the political inability to choose the prudent path and to regulate the next year's catch on the basis of the pessimistic assumptions in the interests of the long-term viability of the stock. The benefit of the doubt has consistently and myopically been given to the exploiter rather than to the stock. Virtually universally the result has been nemesis for the stock where there has been open access.

At the last meeting of the CCAMLR in September 1984, the signs were that the same sad story was repeating itself at least in regard to fin fish. Because of a lack of data about what had happened, for example, to the South Georgia fishery from when it started in 1969 to the entry into force of the Convention, the scientific committee could not develop a fully reasoned series of species catch limits. What data there were pointed to the need for very low catch limits or, more prudently, a total closure of the fishery. But in the absence of sufficient data the margins of error were inevitably large and, just as inevitably, the benefit of the doubt was claimed by exploiters to serve their immediate interests in getting some fish this year even if that meant the probability of getting less in the following year. It has often been asserted that CCAMLR is flawed by the consensus rule under which it makes its decisions and that a majority voting rule would have been more effective. I do not believe this is the case. Decisions in such organisations as CCAMLR are means of registering points that have been reached in what is, in fact, a continuous debate. The means by which such decisions are reached is neutral in regard to their effectiveness. What has eluded all similar fishery conventions, whether government by consensus or majority voting, is the political will to have regard to long-term rather than short-term interests. It is possible that CCAMLR may provide a means of ensuring the long-term interest in regard to krill where the exploitative interests are, for the present, very uncertain.

Future uses of Antarctica

Let us now turn to the future possible activities of mining and using icebergs for water. What problems are likely to arise from these activities?. I will first deal quickly with the use of icebergs for water. If the problems of towing icebergs and of distributing melt water to consumers can be overcome, I do not foresee any great problems arising in the Antarctic from the removal of a few icebergs. It is difficult to foresee the removal of ice at a rate which would have a perceptable environmental effect in the Antarctic; more especially as bergs will be allowed to drift as far north as possible before being put under tow. The problems are more likely to occur at the receiving end where the cooling effect of a melting iceberg could have considerable local environmental impact.

The possibility of mining in the Antarctic is looked upon by many as an ultimate disaster, as some sort of dreadful obscenity. I would number myself as one who greatly hopes that mining will not take place in Antarctica and I have a reasonable expectation that, as long as things are so arranged as to ensure that the profit motive shall operate without interference with respect to the exploitation of Antarctic minerals, no mining will take place. This expectation is based on the assumption that Antarctic minerals are likely to be the world's most expensive to exploit and that product substitution will be profitable at a lower price. But what if I am wrong? It is against that possibility that the Antarctic Treaty Consultative Parties are currently engaged in the negotiation of an Antarctic minerals regime.

There are two problems which I see arising from the possibility of there being economically exploitable minerals in the Antarctic. These, put at their starkest, are:

    i) if Antarctic minerals were to be looked upon as an open-access resource like Antarctic living resources then we could be faced with similar problems of regulation after the event as we are in fisheries;

    ii) assuming that the problem above can be overcome and that regulations can be put in place before activity starts, how can we ensure that those regulations are the right ones and that they are observed?

I do not in fact believe that the first problem (that of whether Antarctic minerals are an open-access resource) is of much consequence. Although there have been many spine-chilling media forecasts of Klondike races to exploit Antarctic minerals and of competition, tension and dispute leading to outright military conflict, such scenarios seem to me to be far out of touch with reality. The huge amounts of money that will necessarily be involved in the exploration and development of Antarctic minerals will require security of title and political stability. No-one in their senses is going to invest that sort of money in a climate of uncertainty and political dispute. It is, moreover, relatively easy to deal with the question of whether Antarctic minerals are an open-access resource by negotiating a regime which says that they are not and that any mineral activity is prohibited until it is permitted. As Arthur Watts will tell you, this is one of the provisions of the regime under negotiation upon which there is already a consensus. Following the conclusion of a regime, it is probable that nothing beyond the prospecting stage will happen until it is agreed that it can happen and the terms and conditions which would apply to it have been fully worked out.

More real are the problems of ensuring that the regulations in place are the right ones and that they are observed. Let us consider these problems from the point of view of environmental protection and remind ourselves of the burden of the non-governmental organisations' challenge, that the Antarctic Treaty Consultative Parties are not capable of enforcing their own rules.

The negotiations are well seised of the problems of compliance, liability and dispute settlement. But whatever conclusions are reached on these vital issues it seems to me difficult to say that there will be no room for legitimate doubt as to whether any international regime for the regulation of Antarctic mineral activity – any international regime that is, not just the one that happens to be under negotiation at the present time between the Consultative Parties – would provide adequate solutions to these problems. There are too many examples of international regulatory activity which represent the lowest common denominator of agreement and of regulations being honoured in the breach rather than in the observance. These are facts of international life and there is little reason to expect that the Antarctic will be exempt from similar ills unless these facts are clearly recognised and specific provisions are made to cope with them. It is, I believe, possible that such provisions could be made but it will require some rather different thinking than is prevalent at the moment. I do not mean by this to undermine the negotiations which are currently afoot. Arthur Watts will tell you why they are necessary and I fully agree with him. But I do think there is a need to reconsider some of the fundamental assumptions on which they are based.

A possible basis for such reconsideration lies in what is politically unique about the Antarctic. The Antarctic is the only place in the world where there is a difference of view as to whether sovereignty can or cannot be exercised; that is to say whether activity is to be regulated by one government or by agreement between many governments. At present the tide of majority opinion, both inside and outside Antarctic Treaty system forums, appears to favour internationalisation at one level or another and, in some quarters, runs strongly against the notion that territorial sovereignty has any relevance in the Antarctic. Time was when I adopted a similar view but the more I have seen of some elements of the operation of the Antarctic Treaty system the more I have come to a feeling of irritation and disenchantment when the ability to arrive at prudent decisions flies in the face of reason. I find that such facts as the following begin to weigh more heavily in the balance:

i) during its first quarter century, Antarctic whaling was adequately regulated by territorial legislation; it was when the activity could be pursued on the high seas beyond the reach of such jurisdiction, using pelagic factory ships, that the decline in stocks began; the two international whaling agreements of 1934 and 1946 did little more than slow down that decline;

ii) competitive exploitation of finite open-access resources puts a premium on rapid recovery of capital investment or, put more colloquially 'If you don't grab the resource while the going is good, some other fellow will';

iii) under such circumstances all that international agreement can achieve is to negotiate a rate of use which lies somewhere between the maximum sustainable yield and a higher rate of use which gives what is asserted to be an adequate return on capital invested;

iv) a state, on the other hand, has a longer-term interest in ensuring that its renewable resources are not used up and can impose the consequences of that interest on exploiters within its jurisdiction.

v) one of the more important motivations behind successive revisions of the law of the sea has been the patent inadequacy of international agreement to prevent the decline of open-access fish stocks, leading to the establishment of coastal state jurisdiction over the majority of such resources and the enforcement of catch limitation.

While I acknowledge that these facts are more obviously relevant to the problems arising from the exploitation of renewable resources, I believe they also offer a basis for some thoughts of relevance to the exploitation of non-renewable resources. The starting point is that the exploiter of any resource, whether it be renewable or non-renewable, living or mineral, has a perfectly clear aim which is to minimise his investment and to maximise his return. His aim is simple and justifiable; there is nothing wrong with it in principle but it needs to be brought up against some sort of regulatory authority if other, long-term interests are also to be secured. Such an authority, to be effective, needs in the final analysis, the power to impose those interests.

Let us now look at the role of a state within whose jurisdiction an exploiter of a mineral resource is operating. Within certain limits imposed by market prices and the ability of an exploiter to react negatively to the conditions made by the state, it is possible for the state to bring other interests to bear on the manner and the rate of exploitation. It can, for example, impose environmental controls in response to public concern and it can impose a depletion policy which is in its interest but which is different from a depletion rate dictated by the exploiters' simple financial interests. In short, a state, as long as it acts reasonably, can impose and enforce other interests on an exploitative venture.

Let us now look at the situation in the Antarctic. In view of the states asserting territorial jurisdiction in Antarctica, they have all the powers necessary for the regulation of mineral activity and the enforcement of their policies. They can, in principle and in fact, be effective. But what about the states not recognising such assertions of territorial sovereignty?. They say that they can be just as effective with regard to their national operators by imposing their policies and jurisdiction on the basis of nationality instead of on the basis of territory. But the difficulty for the non-claimants is that they cannot, as can a territorially sovereign state, give an exclusive title to the resources their operator wishes to exploit. It is this lacuna in the capacity of non-claimants which, in the Antarctic context, gives rise to the necessity for an international agreement such as that which is currently being negotiated.

The difficulties arising from an attempt to regulate Antarctic minerals activity by international agreement are inherent in all such attempts whether in the Antarctic or elsewhere. Starting from policies which the reasonable man would see as being such as ought to be agreed, the progress of negotiation tends towards the lowest common denominator of that which can be agreed in the circumstances of different interests and the time available. In that process an array of interests are likely to be brought to bear, whether openly or not, which have little and sometimes no relevance to the questions to be decided. This sort of situation tends to give a very considerable advantage to anyone in the negotiations who, like a mineral operator, has a clear and simple view of his own interests. Coupled with this diffuseness in policy making goes, hand in hand, a tendency towards diffuseness in the exercise of responsibility in the sense of enforcement capacity.

Let me give an example of the sort of problem which I can foresee arising: a question at issue might relate to the means to be used for transporting offshore oil out of Antarctica. The choices might be between using existing surface tankers, building submarine tankers and constructing a pipeline to a shore-based tank farm. Surface tankers might be environmentally riskier but would be the least expensive; the pipeline might be environmentally safest but would be the most expensive; the technology and associated environmental risks related to submarine tankers is uncertain. Let us suppose that the country from which the operator comes cannot itself provide any of these means of transportation, and that there is competition between countries as to which means should be purchased by the operator. It is not hard to foresee the sort of debate that would ensue, the arm twisting and the hidden agendas which would be operating, or the manner in which the responsibility for the final decisions would be diffused between all the negotiators.

I began by suggesting that some new thinking was needed and that this should derive from that political characteristic of Antarctica which was unique; that is to say that it is the only place in the world where the relative merits of decision making on the basis of territorial sovereignty and on the basis of international agreement can be assessed and the best of both worlds can, if the political will is there, be adopted.

I want to leave you with this thought: if what is desired is to regulate Antarctic resource activity in a responsible and rational manner then there are valid attributes of territorial sovereignty which have a role to play just as much as the valid attributes of international agreement. Those who cry either that territorial sovereignty in Antarctica is an irrelevant neo-colonialist hangover or that the wider the participation in international decision making the better are the decisions reached, are, I believe, allowing themselves to be deceived by conventional rhetoric and are not taking into account the underlying realities. If the claimant case is looked upon as being self-interested, it is a delusion to look upon the case of non-claimants as being without a similar degree of self-interest. What is unique about Antarctica is that it offers an opportunity for a negotiation between such interests. For such a negotiation to take place there must be a recognition that neither set of interests are inherently or morally superior to the other. The grounds for such a negotiation in relation to Antarctic minerals exist; it remains to be seen whether the parties to it can grab the opportunity. I am not without hope.

20

Antarctica: the claims of 'expertise' versus 'interest'

ZAIN-AZRAAI

'Realities' in dealing with Antarctica

Central to all considerations of Antarctica is the fact that there is no agreement on the issue of sovereignty. But it could also be said that except *inter se* the seven claimant states (and leaving aside the matter of overlapping claims), all other states do not recognize these claims. There is also a large unclaimed sector. In addition, two states assert they have a 'basis of claim'. Thus, in relation to the sovereignty issue, there is the reality of the claims, on which the claimant states are insistent; there is also the reality that these claims are not recognised by an overwhelming majority of states.

There are other realities which include the following:

   i) the Antarctic Treaty has existed since 1959 and today has 32 signatory states; a number of conventions and agreements have evolved within the framework of the Treaty dealing with various activities in Antarctica; the parties to the Antarctic Treaty System (ATS) are anxious that its achievements which are based on the fragile agreement of Article IV of the Treaty, should not be prejudiced;

  ii) Antarctica is not a minute atoll of no significance but it occupies some 1/10th of the surface of the globe with a strategic location, fragile ecosystem, rich marine and, possibly, mineral resources; it therefore has great significance for international peace and security, economy, environment, scientific research, meteorology, telecommunications and so on; it also has no permanent human habitation;

 iii) non-Treaty states (NTS) who are excluded from any involvement in the management of Antarctica regard the Antarctic Treaty system as deficient; more specifically they are critical of the rights

211

which the Antarctic Treaty Consultative Parties (ATCP) exercise
within the system, and in light of the sovereignty issue, they have
raised the questions: who should have the right to manage
Antarctica? on what basis? and by what authority? They are also
concerned with the current negotiations on a minerals regime
from which they are excluded.

In the light of these realities, what should be the system for managing
Antarctica which would take into account the interests of all parties?

The 'rights' of the Treaty Parties

A number of the non-Treaty states have been critical of many
features of the ATS. These include the facts that, under the Treaty, the
ATCPs – and they alone – have the rights to make decisions ('exclusive'),
and that the ATS assert rights to regulate all activities in Antarctica
('total'), and that decisions within the ATS are not subject to review or
even discussion by any other body ('unaccountable'). What is asserted are
not special rights such as weighted voting or rights limited to specific
aspects of management in Antarctica but rights which are exclusive, total
and unaccountable. These rights, moreover, are self-conferred on the basis
of a self-determined criteria. Put baldly in this way, such an assertion of
rights by a group of states is surely quite extraordinary and has to be
justified on quite extraordinary grounds, particularly in the context of
contemporary international society which may claim the increasing democratization in international life as its contribution to the evolution of the
management of international affairs.

The ATS case: a right based on 'expertise'

What is the justification for the 'rights' of the Treaty Parties? To
the extent that I can judge it, the extraordinary rights which the Treaty
Parties assert, or rather, bestow on themselves, is based on the argument
of expertise and experience. The management of Antarctica, it is said, is
sophisticated stuff and only those states which have real knowledge or
'expertise' based on actual activity in Antarctica, should have the right
to make decisions relating to it. By the same token, it is obvious that these
states cannot be answerable to that vast majority in the international
community who have no (or insufficient) knowledge or experience of
Antarctica. The essential criteria then, is 'expertise'. No doubt the
argument can be phrased more elegantly but even if I have been somewhat
simplistic or even presumptious in setting out the case of the Treaty Parties,
I cannot see that it can rest on any other argument.

There are, admittedly, supplementary arguments, but it is difficult to see that they can justify the assertion of rights which are (to repeat) exclusive, total and unaccountable. These arguments include:

i) that the current system 'works' (this argument, in turn, opens itself to the view that it would require sanction from a higher authority which could subject its functioning to review in order to ensure that it works better);

ii) that the ATCPs act as trustees for the international community (however trustees, cannot be self-appointed, they should have no material interest in the trust property, and they must be accountable);

iii) that the Non-Consultative Parties (NCPs) and, to a lesser extent, some international organizations, upon invitation, participate in the meetings of the ATCPs (however, such participation, which appeared to have been extracted as reluctant concessions after 23 years of the life of the Treaty, have certain limitations; indeed the very logic of the ATCP case requires a two-tier system where the decision-making rights under the Treaty must rest with the ATCPs);

iv) that the ATS does not confer rights but only imposes obligations (this is slightly disingenuous: all laws and regulations are almost by definition the imposition of obligation while deciding on what the laws and regulations should be is exercising a right; thus, what is the current negotiation regarding the minerals convention other than the exercise of a right to decide on all aspects of mining in Antarctica including whether it should take place, on what terms, and for whose benefits?);

v) that the 'obligation' assumed by the Treaty Parties does not affect the freedom of members of the international community. (Again, in a limited legal sense, this is true almost by definition. But again, to take the example of minerals, if the United Nations – or if the 127 members who are not involved even indirectly in the current negotiations – should adopt a resolution to begin negotiations on a minerals regime, what would be the reaction of the Treaty Parties? Thus the freedom of the NTPs is, in fact, curtailed).

The response of the non-Treaty Parties

The response of non-Treaty Parties (NTPs) to an assertion of this 'right' by Treaty Parties has taken one of two forms. The first consists of denying that, in the contemporary world, there exists or can exist any such right unless it is conferred by the international community. The states which take this position assert that the days when rights might have been asserted on grounds of discovery, occupation, contiguity, inherited right, geological affinity, possession or 'expertise' are long past. They go on to point to the special characteristics of Antarctica (no agreement on sovereignty, lack of permanent human habitation, significance to international society etc.). They then conclude that Antarctica must therefore be the common heritage of mankind to be governed by an international regime duly constituted by the international community. In this sense, their case is as logically self-contained and complete as that of the ATCPs in asserting rights based on 'expertise'. But it comes up against some of the 'realities' of Antarctica (to which reference has been made above).

The other response consists of accepting these 'realities' but goes on to ask: who gave the ATCPs the rights they assert and by what authority? What is the justification for such rights based on expertise? Is this notion acceptable to the international community today? If not, how best should the international community proceed to discuss and evolve a system which would be more generally acceptable?

An approach based on 'interest'

This second response takes the approach not so much of 'expertise' as of 'interest'. This approach begins from the basic question: does mankind as a whole, have a legitimate interest in Antarctica? If so, what should be the objectives of a regime which would best serve mankind's interest? What should be the characteristics of a regime which would best achieve those objectives? To bring the discussion to a more concrete level, does the present Antarctic Treaty system succeed in doing so? If there are deficiencies, what are the possible remedies? States which take this approach respond to these questions as follows.

First, they assert that, bearing in mind the special characteristics of Antarctica, in particular its great significance for international peace and security, economy, environment, scientific research, meteorology, telecommunications and so on, developments in Antarctica are clearly matters of global interest and fall within the ambit of global concern.

Second, they assert that a regime on Antarctica should, among other things, preserve international peace and security, it should promote and facilitate scientific research and exchange, it should protect the environment and it should ensure that the fruits of any exploitation of Antarctica's resources (if there is agreement on such exploitation) should be equitably shared by mankind. Further, they assert that such a regime, serving these objectives, should be one in which member states of the United Nations, as well as the relevant specialized agencies and other international organizations, are appropriately involved, and the regime should be accountable to the United Nations as the most universal and representative international organization.

Third, in testing the current regime of the Antarctic Treaty System against these criteria, they recognize its many practical virtues (non-militarization and non-nuclearization, on-site inspection, promotion of scientific research and, perhaps with some qualification, protection of the environment etc.). At the same time they assert that a system of self-conferred rights which are, moreover, exclusive, total and unaccountable, is seriously flawed and unjustified and it should be remedied. They believe that the concept of the involvement of, and accountability to, the international community regarding the management of Antarctica should be explored further. Specifically with regard to the minerals negotiation, they simply cannot see how these should be the exclusive prerogative of the ATCPs (with NCPs at last as observers) any more than the seabed negotiations could have been confined to a limited group of states.

Fourth, (in considering possible remedies) they assert that they have an open mind. They are conscious of the legal and practical realities. They are not wedded to any particular structure for an acceptable regime. They believe that there should first be agreement on the objectives and the general characteristics of a regime, on the basis of which an acceptable structure should be worked out.

From the above it is clear that there are differing views on how Antarctica should be managed. The question therefore arises: how best might these differences be examined and resolved?

'Expertise' v. 'interest'

Before turning to this question, the issues stated thus far may be examined further by considering the criterion of expertise based on activity, which serves as the essential justification for the rights that the ATCPs enjoy.

The fact is that in the world today, it is normal for states to participate in discussing and dealing with all international issues such as peace and security, disarmament, international trade and finance, decolonization etc. This involvement is justified in terms not of 'expertise' but of 'interest' (i.e. that these issues affect or concern them). International society has in fact moved from the implicit acceptance that 'might is right' to a greater democratization in the management of its affairs. To insist on the principle that 'expertise is right' surely flies against this development. While such democratization is by no means absolute, why should this conceptual progress in the management of international affairs not be extended to Antarctica?

In asserting this, one is not necessarily insisting on a universalist principle (one country, one vote) at all times. The fact is that there are many variations of this principle in a number of important international institutions. Examples include the Security Council where the veto exists, the World Bank and the International Monetary Fund, which has weighted voting, the International Seabed Authority and various commodity arrangements etc. where by mutual agreement and interests of specific groups – producers, consumers and other parties directly affected – are accommodated in many different ways. This takes us into the area of 'special' rights as distinct from 'exclusive' rights. Accepting, without necessarily admitting, that some special expertise is required to make decisions affecting peace and security, the environment and scientific research in Antarctica, should it be exercised exclusively by the ATCPs? And what is the justification for extending this right to cover every aspect of activity in Antarctica including the possible exploration and exploitation of its mineral resources which involve, but is not limited to, the question of the equitable sharing of its benefits, on which the expertise – let alone the exclusive expertise – of the ATCPs is not self-evident? And finally, what is the justification for the exercise of such rights in a manner unaccountable to the rest of the international community?

The next steps
There appear, then, to be essentially two different approaches in dealing with the question of how Antarctica should be managed, one based on 'expertise' and another based on 'interest'. The question, therefore is: how should the international community proceed in attempting to reconcile these different approaches?

One suggestion, which is advanced by the ATCPs, may be paraphrased as follows: join the Antarctic Treaty and seek to remedy whatever short-coming which may exist from within. The essential deficiency of this suggestion is that it is a sort of 'catch-22' situation: it is difficult to see how the NTPs can agree to this suggestion given their view that the present system is conceptually flawed.

The NTPs in turn suggest that in a situation where such differences exist, the normal way to proceed is to establish a forum where the interested participants would be on an equal footing, which would examine in depth all the outstanding issues and consider ways and means in which these differences can be reconciled. Thus, at the 39th session of the United Nations General Assembly, they put forward a proposal for the formation of a United Nations Ad Hoc Committee. Unfortunately, this was firmly resisted by the ATCPs despite assurances that the establishment of such a Committee would be without prejudice to the position of any of the parties and was not intended to be a parallel system (and, indeed, could not have been, given its terms of reference), or to be the 'thin edge of the wedge' to supplant or replace the current system. Bearing in mind that the present machinery under the Antarctic Treaty system is concerned with the practical management of activity in Antarctica, it would appear that such a Committee would be an appropriate forum in which the differences between the ATCPs and the NTPs could be discussed including the following questions:

- what special or, even, exclusive rights should be assigned to the states on the basis of their activity in Antarctica?
- what should be the right of other states?
- how should the notion of 'accountability' be translated into practice?
- similarly, how should the notion of 'involvement' of states other than the ATCPs and of the relevant specialized agencies be translated into practice?
- how should the issue of minerals be dealt with in a way which meets the interests of all parties concerned?
- how can the views of the NTPs regarding the deficiencies of the Antarctic Treaty system be accommodated without prejudicing its achievements?

These are surely serious questions which deserve to be dealt with in a serious way. They are also difficult questions which cannot be resolved by ignoring the complexities of the existing situation. Neither a simplistic attitude unrelated to realities nor an adamant attitude that the experts know best – nor, I may add, a display of power – would be helpful. What is called for is an open-minded approach to reconcile the legitimate interests of all parties in the management of Antarctica which takes into account existing realities.

21

Whither Antarctica? Alternative strategies

J. R. ROWLAND

I have had some connection with several Antarctic Treaty Consultative Meetings, but none with Australian government policy since 1983. Also, the topic I have been assigned is 'a third strategy' – which I have amended to 'alternative strategies' – and I am focusing on that brief. So what I shall write in no way reflects any official Australian position; indeed a good deal may depart from it.

Hegel pointed out the difference between *sollen* and *sein*, between the world as it ought to be and the world as it is in hard reality. The distinction applies forcibly to the case of Antarctica. There are many views about what ought to happen there: what will happen there in fact will be shaped by the interplay of pressures and interests expressed in the policies of governments – with the international conservation movement, in particular, as chorus. It is likely to work itself out over an extended period, and both process and outcome are really impossible to predict.

If I had to make a prediction, my first point would be that the Treaty partners, if they continue to stick together as effectively as they have done so far, and given their considerable recent accretion of membership, are a formidable group. The Treaty has on occasion been described as a house of cards, which could rapidly collapse under pressure. How far this idea is from the truth, and how far the Treaty corresponds to the wide variety of interests on the part of its signatories, is indicated by the way in which all of them, non-Consultative Parties as well as Consultative Parties, have so far given every sign of resistance to change in the Treaty and of support for what has become known as the Treaty system. Including as they do the two superpowers, all the permanent members of the Security Council, China, India, and other substantial Third World countries, together with newcomers like Sweden, Finland, Cuba and Hungary, no change could be made effective without their agreement.

Moreover, the Treaty system has not only attracted new support, but is being adapted and modified in ways which acknowledge current pressures, so that it is not a static target. And, so far, many of the countries from which the chief pressures for change might be expected to come have not shown a great deal of interest. It is therefore possible that, just as the campaign for a New International Economic Order swelled and then sank again, so demands for radical change in Antarctica might also decline.

Nevertheless, Antarctica is back on the international agenda, from which the Treaty removed it for nearly 25 years. The enlargement of the Treaty group itself has changed and will change the situation, in ways yet to be seen. The old Antarctica club – which really did produce a strong sense of solidarity and cooperation, as anyone with experience of it knows – has, it seems, gone forever, though we may hope that something of its spirit will remain. Movement is already under way, and I believe that it will continue.

Essentially, the achievements of the Treaty were three: to reserve Antarctica permanently for peaceful purposes by its demilitarisation and denuclearisation; to ensure freedom of science and scientific information; and to freeze the vexed questions of claims. Originally, conservation was not an issue, but it has attracted increasing attention since, both within the Treaty group and outside it. And, of course, the Treaty did not deal with resources.

About 10 years ago, the Treaty partners began turning their attention to that question. They first negotiated the Convention on the Conservation of Antarctic Marine Living Resources (CCAMLR) to cover living resources, and then attacked the task of negotiating a regime for non-living resources, that is, minerals. CCAMLR has attracted relatively little notice, though some criticism from conservationists for its ineffectiveness in regulating catch, and from others because of its unresolved relationship with United Nations Conference on the Law of the Sea (UNCLOS) and its extension above 60° S – the limit of the Treaty area – to a line defined as representing the Antarctic Convergence. But the minerals negotiations focussed international attention on Antarctica, and they are the key to the current pressures and criticisms.

These criticisms, and the answers made to them by the Treaty partners, have been described by other speakers. I shall not go over that ground again, except to say that:
  i) the essential charge is that the Treaty partners are a limited and self-appointed group which has no mandate to legislate for what should be the 'common heritage of mankind', and which is also undemocratic in that decision-making is reserved to those who have passed the Treaty's test of Antarctic activity; and that
 ii) there seems to be general acknowledgement that the permanent

reservation of Antarctica for peaceful purposes, its demilitarisation and denuclearisation, and freedom of science and scientific information should be maintained; also that the importance of conservation should be recognised.

Let me add three general points. First, there can be no stable and lasting solution for Antarctica which is not a consensus solution. That means – second – that it must be acceptable to each of the four interest-groups involved, defined by Keith Brennan, to whom I am indebted for this analysis, as:

i) those who assert sovereignty, henceforth referred to as the claimants,

ii) those who regard themselves as having a 'basis of claim', i.e. the United States and the USSR,

iii) the Antarctic Treaty partners who recognise neither claims nor basis of claim, and

iv) those outside the Treaty who assert the 'common heritage' doctrine and the need either for a completely new regime or, at least, for the Treaty powers to accept more formal responsibility to the international community.

To those four groups, which are groups of states, I would add the international conservation movement, which is beginning to make itself felt on national governments.

Third, a consensus solution implies the need in the long run for some kind of universality or, at least, for a framework which does not exclude any substantial number of countries not bound by it or capable or threatening it. This may seem to belong to the world of *sollen*, but it does have practical implications, as I shall explain in a moment. In the long run – and however it is done – there needs to be some more effective marriage between the Treaty system and the principle of universality than at present.

There seem to be two paths forward. One has already been embarked on; it is that of the evolution of the Treaty system, in ways which will attract new members to it – and which will therefore have to offer inducements to membership by 'democratising' it. This implies increasing the role of the non-Consultative Parties, diminishing the difference between them and the Consultative Parties at Consultative Meetings, and probably making it easier for countries to attain consultative status by a more flexible interpretation of the necessary qualifications. It also implies greater recognition of the interests of the world community by increasing the flow of information from meetings taking place under the Treaty system, and strengthening the link with the United Nations, through the attendance of representatives of Specialised Agencies at such meetings, through providing their reports to the United Nations Secretary-General for circulation to United Nations members, and so on. Whether there should be continuing discussion of such reports at the United Nations, notably at the General Assembly, is a further question. The present position of the Treaty partners is clearly against any formal accountability, just as they would oppose the creation of new institutions in the United Nations framework.

Finally, it could mean something more to deal with conservation. There are non-governmental suggestions, for example, for a conservation convention within the Treaty system – perhaps even along the lines of CCAMLR, so that it could include non-signatories of the Treaty and have its own central body.

Negotiations to widen participation within the Treaty system are proceeding. It is a perfectly possible way forward; and because it is the way the Treaty partners have chosen, it is in practical terms the most promising. Let me speculate briefly about this path, the evolution of the Treaty system.

First, all 32 Treaty partners will be able to attend future regular Consultative Meetings. As to Special Consultative Meetings, the position is not yet so clear, but the fact that all signatories are now able to attend the meetings on the central question of minerals surely implies that they are most unlikely to be excluded from other Special Consultative Meetings in future. They are attending, of course, as observers. But once present, with the right to speak and circulate papers, a country can make itself felt, even if it is not counted in the final decision-making. The tradition of compromise and consensus should have its effect here; the chairman of the meeting and the Consultative Parties themselves will be reluctant to override a substantial body of opposition among the non-Consultative members.

But the manner of operation of a group of 32 is likely to become rather different from that built up over the years among the smaller group. So far, it seems, a good deal of the old atmosphere remains and the new members have not been particularly active or vocal. But will this continue? Will there be the development of lobbying, the formation of groups, and the other phenomena of a normal international conference? How will the tradition of consensus fare? Will the presence of South Africa prove an irritant which will cause countries to take positions they have not felt obliged to take so far? (On this point, I would only remark that South Africa has so far a good record of low-key and constructive behaviour within the Treaty, and that, since it is a country active in Antarctica, it is surely better that it should be included rather than excluded.) Among the Consultative Parties, what role will be played by the newcomers, notably China, India and Brazil?

Another point is that the balance between claimants and non-claimants will be different. Seven claimants out of a total body of 32 is not the same as 7 out of 12 or 16. Will this have an effect? I am not sure how far it will lead to changes, at least in the short term. Adherence to the Treaty implies no recognition of claims, but it does imply recognition that there are claimants. And this is more important than it may sound, especially given the importance of the countries concerned, the consensus rule, and the practical impossibility of solutions which the claimants would unite to oppose. I shall say something more about the claims in a moment.

A variant of the evolutionary path which should be mentioned is that indicated in Article XII of the Treaty; namely, revision of the Treaty by unanimous agreement among Consultative Parties, or a review conference, which could be called from 1991 onwards at the request of a Consultative Party – though any amendment would require a majority of Treaty members, including a majority of Consultative Parties. So far, there seems no prospect whatever of the first, and very little indeed of the second.

Let us now consider the second and more radical path – which the Treaty partners have rejected. It is that of new institutional arrangements, necessarily requiring renegotiation of the Treaty, and probably accountability to some worldwide body – in particular, the United Nations. Here I can say only that it is possible to imagine various models, all at present equally theoretical, and to mention some of those which have been put forward.

One would be a trusteeship over Antarctica by the present Antarctic Treaty powers, presumably reporting to the United Nations, as the trustee powers used to report to the Fourth Committee. This was originally suggested by the Malaysian Prime Minister, Dr Mahathir, in 1982, though he did not spell it out in detail, and I do not think he has returned to the idea.

A second would be that proposed by Antigua-Barbuda: to find ways by which the Antarctic Treaty could be modified:

  i) to accommodate the principle of universality in terms of accession to the Treaty (but the Treaty partners would say that the Treaty is already open to all); and

 ii) to establish a system by which the supreme decision-making body of Antarctica is made up of the existing Consultative Parties as permanent members and representatives of regions as non-permanent members reporting to the General Assembly of the United Nations. Non-governmental environmental organisations would attend all meetings of the Antarctic authority, which would establish and administer a 'system of international taxation and revenue-sharing', which would cover both living and non-living resources.

A third and more radical proposal is that of Pakistan, which calls for 'a (new) international regime freely negotiated between the members of the international community under the auspices of the United Nations' and 'established by an international treaty of a universal character', providing for the 'development and management of Antarctica and an equitable sharing by States of its resources...taking into particular consideration the needs and interests of the developing countries'. This regime would preserve the existing Treaty principles of demilitarisation, denuclearisation, and freedom of science, adding environmental protection and various other points stemming from the principle that Antarctica is 'the common heritage of mankind' and that no state can 'claim or exercise sovereignty' there. This proposal marks perhaps, the present extreme of *sollen* as opposed to *sein*.

A fourth proposal would be that made by E. Luard (who worked with the Department of Foreign Affairs in the summer of 1984) for hiving off the minerals question, leaving the present Treaty in place to deal with demilitarisation, science, and (he adds) the environment. Sri Lanka made a somewhat similar suggestion in the 39th Session of the General Assembly. To deal with minerals, Luard suggests a new body of about 30, with special representation for those countries with particular interests such as claimants, geographical neighbours, potential exploiters and those with special historical links. Such countries would evidently be permanent members, but would be balanced by developing countries – Brazil, India and China are mentioned – plus others chosen on a regional basis. Luard cites the International Labour Organisations (ILO), International Aviation Organisation (ICAO), International Monetary Fund (IMF) – and indeed the Council set up under UNCLOS – as offering a broad precedent. (The International Atomic Energy Agency (IAEA), with its permanent seats on the Board of Governors for those 'most advanced in the technology of atomic energy' in various regions, would provide a similar model.) He does not, however, suggest that this body should report to some kind of wider general conference, or to the United Nations.

Could the problem be divided in this way? Certainly, the Treaty partners have so far created no central organisation beyond the rudimentary secretariat arrangement to look after matters between consultative meetings; the Treaty is still no more than a network of obligations among its parties. CCAMLR has created an organisation, with its own membership.

And the present minerals negotiations, of course, do envisage a separate organisation.

Like CCAMLR, it would be based on the Treaty and would be an extension of the Treaty system. It would require acceptance by its members, if not of the Treaty itself, then of various basic provisions of the Treaty and various instruments established under it, and acknowledgement of the special responsibilities of the Consultative Parties. It would be intended to bring in all states or parties which might engage in minerals activity in Antarctica. But it would not introduce a new principle, such as that of regional representation by countries whose function would be to represent the international community. Minerals being at the heart of the matter, it is hard to see the Consultative Parties accepting a new principle that could make the rest of the system peripheral.

These, then are all alternative strategies. However the debate may work itself out, I should like to add a few comments. First – and this is an observation rather than a proposal, though I see merit in it in terms of universality – I think that the United Nations General Assembly would endorse without difficulty the permanent reservation of Antarctica for peaceful purposes, its demilitarisation and denuclearisation, freedom of science and inspection, and so on (including the importance of conservation). Not all these matters have been mentioned by all speakers, but the general trend of opinion in the Assembly seems clear enough.

Whether the Assembly would endorse Article IV – the freezing of claims – is of course more doubtful. Article IV is interpreted by the advocates of common heritage as incompatible with that principle, no doubt because, as I have said, it implies recognition, if not of claims, then of the fact that there are claimants. The claims are regarded as 'outmoded'. It is believed that they should be swept aside and a new start made. But the claims are a legal and political fact, and recently the tide seems to have been running in favour of greater rather than less assertiveness about them. No one, and certainly no UN resolution, can force a country to abandon a claim; and the principle that there can be no solution without consensus would still seem to require acknowledgement that claims exist. One strong reason why the Treaty partners defend the Treaty is precisely the desire to keep the genie of disputes over sovereignty within the bottle of Article IV. In the long run, the claims may become submerged, in the course of some process of bargaining, in return for some advantage of acknowledged status, but that day has not yet come. One may at least pose the question whether, if the non-recognisers within the Treaty can accept Article IV, those outside it could not do so.

Second, something more needs to be said about minerals, for this is the crux of the problem. I am uneasy about the present negotiations, because I believe that any regime worked out between the present Treaty partners – for now all of them, not only the Consultative Parties, are involved – and then offered to the outside world for adherence, risks instability. (It will be difficult enough to reach agreement among the present group; notably, the 'bifocalism' which enabled the negotiation of CCAMLR will be much harder to adapt to minerals, though the Beeby drafts offer ingenious solutions.) Certainly, the participation of important new members, including India, China, and Brazil, not to mention the Non-Consultative Parties now invited, together with the custom of consensus, will affect the degree of attention to outside interests which any regime will have to give – what is called the 'external accommodation'. Even so, the danger remains. For example, no Arab state and no member of the Organisation of Petroleum Exporting Countries (OPEC) is now a member of the Treaty. If oil were discovered – and that is really what we are talking about – could one not imagine a situation in which Arab (or multinational) finance, let us say, organised an exploiting consortium based in Liberia? The drafts contain provisions directed towards discouraging activity by outsiders, but what sanctions would, in fact, be possible to prevent such a consortium ignoring a regime to which its participants were not parties?

The negotiations are now under way, and presumably we must await their results. I would say only that, if they should run into intractable problems, the time-honoured Antarctic principle of postponement might be helpful. But it would now need to be expressed in something that would itself be a minerals regime. Specifically, a 'freeze' on mineral resources, comparable to that on claims, could be achieved by declaring Antarctica, perhaps not a World park, but a World Conservation Area or some similar term.

That would leave Antarctic mineral resources available, if they should be required in future. It could disarm the advocates of common heritage, who would, I think, find it much less easy to attack than a regime which must inevitably arouse expectations of exploitation by creating mechanisms to regulate it. It would avoid the threat to the Treaty which a breakdown in negotiations might create. It would suit the conservationists. It would avoid the problem of claims, and might even suit the claimants. But it would, of course, require the consent of those whose interest lies in exploitation.

The key countries in this respect are probably Japan and the United States, though I do not underestimate the interest being shown by others. The latter has already made clear that unless a minerals regime is negotiated it will not consider itself bound by the present moratorium. So the attributes of governments are, as usual, crucial. They, in turn, will be much affected by the state of technology, which will determine the likely practicability and profitability of Antarctic mineral exploitation. One view is that exploitation is still a very distant prospect, another that it is not. There is merit in the view that a regime is best negotiated before pressure for exploitation arises, but the objection remains that it should cover at least all potential exploiters. A World Conservation Area, preferably with some kind of United Nations or United Nations Environment Programme (UNEP) endorsement in order to universalise it, would remove the pressure and could offer such a regime.

If, however, the present negotiations prove successful, I shall conclude by repeating that for any regime to be stable, it will need to be acceptable to all the interest groups concerned. I shall also express the hope, whether from the world of *sollen* or that of *sein*, that a continued moratorium, within the regime, on the exploration and exploitation – and even on prospecting, so far as that can be separated from science – will be possible. In practical terms, that would mean at least the necessity of a consensus decision on these matters, and much emphasis meanwhile on the importance of environmental considerations and of the fullest scientific study. However large Antarctica may appear, we simply do not know enough about the possible effects of exploitation there to embark on it without a great deal more knowledge than we now possess.

Part VI
Conclusion

22

The United Nations in Antarctica? A watching brief

GILLIAN D. TRIGGS

Internal resolution of the issue of sovereignty in a minerals regime has no legal effect upon the rights and interests of the international community. Diplomats, scientists and government officials are now more ready to concede that some form of external accommodation must be made. Assertions of common heritage over Antarctic resources cannot be ignored, if only because of the power of the United Nations (UN) to develop a rival regime creating international discord and rendering unworkable any regime adopted by the Consultative Parties.

The notion that the United Nations should have some role in Antarctica is of long standing. Proposals for United Nations involvement were made after the Second World War and again by India in 1956. It was not, however, until September 1982 that the Malaysian Prime Minister, Dr Mahathir, raised the issue in a speech at the United Nations General Assembly in which he called for a meeting to 'define the problem of these uninhabited lands'. He alleged that the Antarctic Treaty was a neocolonial document which 'does not reflect the true feelings of the members of the United Nations'. He suggested further that all claimants to Antarctic territorial sovereignty 'must give up their claims so that either the United Nations can administer these lands or the present occupants act as trustees for the nations of the world'.

In the following year, Malaysia successfully argued for Antarctica to be included on the agenda for the General Assembly. After debate, a resolution was adopted calling upon the Secretary-General to conduct a comprehensive, factual and objective study on Antarctica and to report back in 1984. The Secretary-General submitted his report to the 39th Session in November 1984, however, as delegations had no time to examine the report it was decided to retain the question of Antarctica on the agenda for the 40th Session in 1985. Nonetheless, Malaysia's attempt to seek the establishment of an *ad hoc* United Nation's Committee on Antarctica was unsuccessful.

The Secretary-General reported that states have various views on the question of a common heritage as it applies in Antarctica. Some argue that, because they accept that territorial sovereignty in Antarctica is valid, common heritage has no relevance in Antarctica; others argue that the concept is a logical extension of an international trend established in outer space and the deep seabed; and others take the position that flaws in the Antarctic Treaty system can be adjusted in an evolutionary way to meet the interests of the international community. It is significant that whilst some states consider that the notion of a common heritage applies in Antarctica they also take the position that it can be given effect within an expanded and reformed treaty. Thus there is no cohesive 'Third World' view which is necessarily juxtaposed to that of the Treaty Parties, and there is no cohesive or clearly defined block view taken in General Assembly debates.

The Question of Antarctica was debated at the 40th session of the General Assembly and three resolutions were passed.

- One, submitted on behalf of the Group of African States, urges Antarctic Treaty Consultative Parties to exclude the racist apartheid regime of South Africa from participation in meetings of Consultative Parties.
- The second resolution requests the Secretary-General to update and expand the present study on the Question of Antarctica, by specifically addressing the question of availability of information from Consultative Party meetings, the involvement of specialised agencies of the United Nations and intergovernmental organisations in the Antarctic Treaty system and the significance of the United Nations Convention on the Law of the Sea in the Southern Ocean.
- The third resolution invites the Consultative Parties to inform the Secretary General of their negotiations to establish a minerals regime and to confirm that any exploitation of Antarctic resources will ensure international peace and security, protection of the Antarctic environment, non-appropriation and conservation of Antarctic resources and 'the international management and equitable sharing of the benefits of such exploitation'.

As to the first resolution, the Consultative Parties are unlikely to expel South Africa from meetings. South Africa's role as a Consultative Party remains, for this reason, a focus for conflict both within the United Nations and the Antarctic Treaty system.

As to the second resolution, the Parties noted at the Thirteenth

Consultative meeting, the need to ensure freer availability of information about the Antarctic Treaty system. They recommended to their governments that, among other things, an Antarctic Treaty handbook should be maintained as a current compilation of all Recommendations and other actions and that final reports should provide full and accurate records of meetings. Consultative Parties have, nonetheless, maintained confidentiality with regard to negotiations for a minerals regime. Such documents as are available have been leaked and are dated. While the reason for confidentiality can be understood as agreement is more likely in the absence of a public debate upon a draft which merely represents negotiating options, a more politically sensitive approach might have been to demonstrate international responsibility for Antarctica within the Treaty system by consulting openly with interested groups and making working documents available.

The second resolution also considers the role of other organisations in the Antarctic system. It reflects the criticism that the Treaty Parties have operated in virtual isolation from other relevant international bodies. The closest links have been between the Consultative Parties and the Scientific Committee on Antarctic Research (SCAR) of the International Council of Scientific Unions (ICSU) – links which grew from the role the ICSU played in developing cooperation between the Consultative Parties during the International Geophysical Year. The ICSU itself has consultative status with the United Nations Economic and Social Council (UNESCO), Economic and Social Council (ECOSOC), Food and Agricultural Organisation (FAO), and working relations with the International Trade Organisation (ITO), World Health Organisation (WHO), and World Meteorological Organisation (WMO). It is these links which demonstrate that, a wide range of interaction exists between the Antarctic Treaty system and the United Nations and other relevant scientific bodies. The Consultative Parties did not consider this issue at the Thirteenth Consultative meeting. However, the General Assembly resolution on this question some days after this meeting should stimulate some response at the next meeting in 1987.

The second resolution concerns the impact of the 1982 Convention on the Law of the Sea in the Southern Oceans (UNCLOS). As described in Part II, the International Seabed Authority has jurisdiction beyond the limits of national jurisdiction which, on one view, includes the deep seabed surrounding Antarctica. While there may have been a silent agreement at UNCLOS negotiations to exclude the Southern Oceans, the Convention itself makes no exclusion of polar regions nor, indeed, does it make any special provision for the regulation of the Arctic or Antarctic.

The third resolution is simply a request for information. Nonetheless, the preambular clauses demonstrate the central purposes of the resolution. They recall declarations adopted by non-aligned countries in New Delhi in 1983, and again in Luanda in 1985, and by the Organisation of African Unity in Addis Ababa in 1985. The resolution records the fact that no states other than Antarctic Treaty Parties are privy to the minerals negotiations, and recognises that the exploration and use of Antarctica should be in the interests *inter alia* of promoting international cooperation for the benefit of mankind as a whole. It remains to be seen how much detailed information the Secretary-General will be given, but the General Assembly has made clear the premises upon which any acceptable minerals regime must be based.

As the General Assembly continues its information-gathering on Antarctica the Consultative Parties are attempting to meet criticism of the Treaty system. At the Thirteenth Meeting, the Parties considered the need for environmental impact assessment of scientific activities and for a revised code of conduct for expeditions. They also considered possible additional measures to protect the Antarctic environment particularly where activities are concentrated in certain areas. It was recommended that governments should invite SCAR to undertake a review of waste-disposal aspects of scientific research and logistical activities. It was also recommended that governments should invite SCAR to offer scientific advice on the system of protected areas in Antarctica and on steps that could be taken to improve the availability of scientific data. It was recommended that governments should cooperate where their stations are established in the same areas to avoid adverse environmental effects arising from their activities.

Finally, the Parties decided that the decision to invite Non-Consultative Parties to the Thirteenth Meeting should now be placed on a permanent basis. There was further discussion on the question of inviting other international organisations to appoint observers to Consultative Party meetings, but this matter was left unresolved. It may be doubted whether these discussions and recommendations quite meet the criticism that decision-making under the Antarctic Treaty System is exclusive, undemocratic and discriminating against small states. There is substance to the argument however, that those who are directly engaged in Antarctic activities are, through their experience in Antarctica, well placed to take decisions which affect matters such as scientific programmes and environmental protection. It is also true of the present Consultative Parties that there is no ideological or economic division on North/South or East/West lines within the Treaty framework.

While the negotiations and recommendations of the Thirteenth

Consultative meeting demonstrate the 'art of the possible', they may be a long way from satisfying the concerns of environmentalists or of political opponents within the General Assembly. It is difficult to speculate upon the future role of the United Nations in Antarctica. When doing so, however, it should be remembered that each of the permanent members of the Security Council is a Consultative Party under the Antarctic Treaty. It has also been noted that the parties represent a high percentage of the world's population and most geographical, economic and ideological interests, with the exception of the states of Africa. Any state may accede to the Antarctic Treaty and its related conventions. These facts suggest that it might be possible to demonstrate responsible management of Antarctica within the Antarctic Treaty system and to defuse demand for 'universal' regulation. For the present, the Secretary General is merely to report, and moves to establish a special UN Committee on Antarctica have thus far been defeated. The UN can be expected to maintain a watching brief on Antarctic activities and to play a closer role in the negotiations upon a minerals regime.

Selected reading

1 *Documents on the Antarctic Treaty System*
W. M. Bush, *Antarctica and International Law* (1982), Oceana Publications, Inc., New York.

2 *Recent collections of essays on legal, political and scientific aspects of Antarctic cooperation and resources*
Francisco Orrego Vicuña (ed.), *Antarctic Resources Policy* (1983), Cambridge University Press.
Rudiger Wolfrum (ed.), *Antarctic Challenge*. (1984), Duncker & Humblot, Berlin.
Philip W. Quigg, *A Pole Apart. The Emerging Issue of Antarctica* (1983), McGraw-Hill.
Jonathan, I. Charney (ed.), *The New Nationalism and the Use of Common Spaces* (1982), ASIL.
F. M. Auburn, *Antarctic Law and Politics* (1982), C. Hurst & Co.

3 *Problems of Antarctic jurisdiction*
Richard B. Bilder, 'Control of criminal conduct in Antarctica', *Virginia Law Review* (1966), Vol. LII, 231–85.
Eric W. Johnson, 'Quick, before its melts: Towards a resolution of the jurisdictional morass in Antarctica' (1976), *Cornell International Law Journal*, **10** (1), 173–98.
Elizabeth K. Hook, 'Criminal jurisdiction in Antarctica' (1978), *University of Miami Law Review*, **33**, 489–514.
John Kish, *The Law of International Spaces* (1973), Sijthoff, pp. 70–81, 116–27.
Richard B. Bilder, 'The present legal and political situation in Antarctica', in Jonathan I. Charney, *op. cit.*, pp. 167–205.
Gillian D. Triggs, 'The Antarctic Treaty regime: a workable compromise or a purgatory of ambiguity' (1985), *Case Western Reserve Journal of International Law*, **17**, 199–228.
Gillian D. Triggs, *International Law and Australian Sovereignty in Antarctica*. (1986), *Legal Books*, Sydney.

4 *Regime for the Antarctic marine living resources*
James N. Barnes, 'The emerging convention on the Conservation of Antarctic Marine Living Resources: an attempt to meet the new realities of resource exploitation in the Southern Ocean', in Jonathan I. Charney, *op cit.*, pp. 239–86.
David M. Edwards & John A. Heap, 'Convention on the Conservation of Antarctic Marine Living Resources: a commentary' (1981), *Polar Record*, **20** (127), 353–62.

Danield Vignes, 'La Convention sur la Conservation de la faune et de la flore marines de l'Antarctique' (1980), *Annuaire Français de Droit International*, 741–72.

Ronald F. Frank, 'The Convention on the conservation of Antarctic Marine Living Resources' (1983–1984), *Ocean Development and International Law*, 13, 291–345.

Rainer Lagoni, 'Convention on the Conservation of Marine Living Resources: a model for the use of a common good?', in Rudiger Wolfrum, *op. cit.*, pp. 93–108.

5 The regime for Antarctic mineral resources

Jonathan I. Charney, 'The future strategies for an Antarctic mineral resource regime – can the environment be protected?', in Jonathan I. Charney, *op. cit.*, pp. 206–38.

F. M. Auburn, *Antarctic Law and Politics* (1982) Chapter 8, Hurst & Co., London.

C. D. Beeby, 'An overview of the problems which should be addressed in the preparation of a regime on Antarctic mineral resources: basic options', in Francisco Orrego Vicuña, *op. cit.*, pp. 191–8.

Francisco Orrego Vicuña, 'The definition of a regime on Antarctic mineral resources: basic options', in Francisco Orrego Vicuña, *op. cit.*, pp. 217–27.

Keith Brennan, 'Criteria for access to the resources of Antarctica: alternatives, procedure and experience applicable', in Francisco Orrego Vicuña, *op. cit.*, pp. 217–27.

Barbara Mitchell, *Frozen Stakes, The future of Antarctic Minerals* (1983), Cambridge University Press.

William E. Westermeyer, *The politics of Mineral Resource Development in Antarctica. Alternative regimes for the future* (1984), Chapter 3. Boulder: Westview Press, Inc. xv, 276.

Rudiger Wolfrum, 'The use of Antarctic non-living resources: the search for a Trustee?', in Rudiger Wolfrum, *op. cit.*, 143–63.

Evan Luard, 'Who owns the Antarctic?', *Foreign Affairs*, Summer 1984, 1175–93.

Rainer Lagoni, 'Antarctica's mineral resources in international law' (1979), ZAORV, 39, 1–37.

Gillian D. Triggs, 'A minerals regime of Antarctica: compromise or conflict' (1984), *Australian Mining and Petroleum Law Yearbook*, 525–54.

Maaster J. de Wit, *Minerals and Mining in Antarctica* (1985), Clarendon Press, Oxford.

6 Antarctica and the Law of the Sea

Alfred van der Essen, 'L'Antarctique et le droit de la mer' (1975–76) *Revue Iranienne de Relations Internationales* (5–6), 89–98.

Alfred van der Essen, 'The application of the law of the sea to the Antarctic continent', in Francisco Orrego Vicuña, *op. cit.*, pp. 231–242.

Francisco Orrego Vicuña & Maria Teresa Infante, 'Le droit de la mer dans l'Antarctique' (1980), *RGDIP*, 340–50.

Francisco Orrego Vicuña, 'The application of the law of the sea and the Exclusive Economic Zone to the Antarctic continent', in Francisco Orrego Vicuña, *op. cit.*, pp. 243–51.

Maria Teresa Infante, 'The continental shelf of Antarctica: legal implications for a regime on mineral resources', in Francisco Orrego Vicuña, *op. cit.*, pp. 253–64.

Ralph L. Harry, 'The Antarctic regime and the Law of the Sea Convention: an Australian view' (1981), *Virginia Journal of International Law*, 21, 727–44.

Christopher C. Joyner (1981), *Virginia Journal of International Law*, 21, 691–725.

S. Muller, 'The import of UNCLOS III on the Antarctic regime'. In Rudiger Wolfrum, *op. cit.*, pp. 169–76.

Appendix 1

UN GENERAL ASSEMBLY RESOLUTIONS
Fortieth session
FIRST COMMITTEE
Agenda item 70

Question of Antarctica
Bangladesh, Brunei Darussalam, Indonesia, Malaysia, Mali, Nigeria, Oman, Pakistan, Rwanda and Sri Lanka: draft resolution

The General Assembly,

Recalling its resolutions 38/77 of 15 December 1983 and 39/152 of 17 December 1984,

Having considered the item entitled 'Question of Antarctica',

Welcoming the increasing international awareness of and interest in Antarctica,

Bearing in mind the Antarctic Treaty and the significance of the system it has developed,

Taking into account the debate on this item at its fortieth session,

Convinced of the advantages of a better knowledge of Antarctica,

Affirming the conviction that, in the interest of all mankind, Antarctica should continue forever to be used exclusively for peaceful purposes and that it should not become the scene or object of international discord,

Recalling the relevant paragraphs of the Economic Declaration adopted at the Seventh Conference of Heads of State of Government of Non-Aligned Countries, held at New Delhi from 7 to 12 March 1983, and of the Final Declaration of the Foreign Ministers of Non-Aligned Countries at their meeting held at Luanda from 4 to 7 September 1985, as well as the resolution on Antarctica adopted by the Assembly of Heads of State and Government of the Organization of African Unity at its meeting held at Addis Ababa from 18 to 20 July 1985,

Conscious of the significance of Antarctica to the international community in terms, *inter alia*, of international peace and security, economy, environment, scientific research and meteorology,

Recognizing therefore the interest of mankind as a whole in Antarctica,

Bearing in mind the coming into force of the United Nations Convention on the Law of the Sea,

Noting once again with appreciation the study on the question of Antarctica (A/39/583),

Convinced that it would be desirable to examine further certain issues affecting Antarctica,

1. *Requests* the Secretary-General to update and expand the study on the question of Antarctica by addressing questions concerning the availability of information from the Antarctic Treaty Consultative Parties to the United Nations, on their respective activities in, and their deliberations regarding Antarctica, the involvement of the relevant specialized agencies and intergovernmental organizations in the Antarctic Treaty System and the significance of the United Nations Convention on the Law of the Sea in the Southern Ocean;

2. *Requests* the Secretary-General to seek the co-operation of all Member States, the relevant specialized agencies, organs, organizations and bodies of the United Nations system, as well as the relevant intergovernmental and non-governmental bodies, in the preparation of the study by inviting them to transmit as appropriate, their views and any information they may wish to provide;

3. *Requests* the Secretary-General to submit the study to the General Assembly at its forty-first session;

4. *Decides* to include in the provisional agenda of its forty-first session the item entitled 'Question of Antarctica'.

Appendix 2. *Approval, as notified to the government of the United States of America, of measures related to the furtherance of the principles and objectives of the Antarctic Treaty*

	First Meeting	Second Meeting	Third Meeting	Fourth Meeting	Fifth Meeting	Sixth Meeting	Seventh Meeting	Eighth Meeting	Ninth Meeting	Tenth Meeting
Recommendations adopted at...	16	10	11	28	9	15	9	14	6	9
Effective date:	30-4-62	11-1-63	27-7-66 (III-I to VI, III-IX and III-X); 1-9-66 (III-XI); 22-12-78 (III-VII)	30-9-68 (IV-20 to IV-28)	26-5-72 (V-1 to 4, and V-9); 31-7-72 (V-7 and 8)	10-10-73 (VI-1 to 7 and VI-11 to 15)	29-5-75 (VII-1,2,3,6,7 and 8); 4-8-83 (VII-4 and VII-9)			
	Approved	Approved	Approved	Approved	Approved	Approved	Approved	Approved	Approved	Approved
Argentina	All 20-9-61	All 11-9-62	All 3-9-65	All 14-12-67	All 12-5-69	All 10-10-73	All 17-10-74	Approved	Approved	Approved
Australia	All 6-10-61	All 18-10-62	All 2-9-64	All 26-2-68 (1) 1-9-80 (11)	All 26-5-72 1-9-80	All 26-5-72[a] 1-9-80	All 33-7-73	All	All	All
Belgium	All 16-2-62	All 13-12-62	All 7-12-64	All 5-12-67 (2)	All 31-7-69	All 8-6-71[b]	All 30-4-73 25-1-78 (6)	All	All	All
Chile	All 19-4-62	All 17-10-62	All 26-2-65	All 21-2-68 (3) 23-11-70 (4)	All 13-4-70	All 3-8-71	All 1-6-73[b] 11-3-75[a] 24-6-81[b]	All	All	All
France	All 6-3-62	All 8-11-62	All 8-11-62	All 30-9-67	All 3-4-69	All 23-4-71	All 11-3-73	All	All	All
Germany, Federal Republic of	All 17-2-81	All 17-2-81	All but 3-8 17-2-81 (1)	All but 4-1 to 4-14 17-2-81 (12) 30-9-68 (5)	All but 5-5, 5-6 17-2-81 (10)	All but 6-8, 6-9, 6-10 17-2-81 (1)	All but 7-4, 7-5, 7-9, 8-9 17-2-81 (7) (9)	All but 8-1, 8-2, 8-5	All	All

Appendix 2. (*cont.*)

	Effective date: 30-4-62	Effective date: 11-1-63	Effective date: 27-7-66 (III-I to VI, III-IX and III-X); 1-9-66 (III-XI), 22-12-78 (III-VII)	Effective date: 30-9-68 (IV-20 to IV-28)	Effective date: 26-5-72 (V-1 to 4, and V-9); 31-7-72 (V-7 and 8)	Effective date: 10-10-73 (VI-1 to 7 and VI-11 to 15)	Effective date: 29-5-75 (VII-1,2,3,6,7 and 8) 4-8-83 (VII-4 and VII-9)		
Japan	All 20-1-62	All 16-11-62	All except VIII 19-1-65 (2) 1-9-66 (4)	All except 1-19 1-11-82 (13)	All except 5 y 6 15-7-69 31-7-72[a] 1-11-82[b]	All except 8,9,10 31-7-72 1-11-82 (7)	All except 8-8-74 1-11-82 (5)	All except 1,2,5 y 9 All	All All
New Zealand	All 4-10-61	All 1-11-62	All 1-12-64 (1)	All 26-6-68 (6) 23-12-71 (7)	All 20-5-69	All 23-12-71	All 26-3-74	All	All
Norway	All 2-3-62	All 11-1-63	All 1-12-65	All 5-7-68	All 18-7-69 19-9-65	All 17-9-71	All 19-10-73	All	
Poland	All 11-7-77	All 11-7-77	All 11-7-77	All 11-7-77	All 11-7-77	All 11-7-77	All 11-7-77	All	
South Africa	All 30-4-62	All 15-11-62	All 5-9-64	All 22-5-67	All 6-5-69	All 21-5-71	All 22-5-73	All	
USSR	All 15-12-61	All 4-10-62	All 20-2-65	All 12-9-68	All 29-12-69	All 4-3-72	All 4-2-74	All	
United Kingdom of Great Britain and Northern Ireland	All 1-12-61	All 30-11-62	All except 4-12-64 (1)	All except 12 11-9-68 (8)	All 8 y 10[a] 27-5-69 11-3-70	All except 5/2 25-6-71	All 29-5-75·	All	
USA	All 12-12-61	All 4-12-62	All 27-7-66 (3) 22-12-78 (5) 31-7-79 (6)	All 23-3-68 (9) 31-7-79 (10)	All 7-6-69 14-10-70[b] 31-7-79	All 25-7-72 31-7-79	All 6-2-73 31-7-79 (5)	All	

[a] 8 and 10 accepted as interim guidelines.
[b] 5 accepted as interim guidelines.